Environment, Energy, and Economic Development Program

# Adapting to a Changing Colorado River

## Making Future Water Deliveries More Reliable Through Robust Management Strategies

David G. Groves, Jordan R. Fischbach, Evan Bloom, Debra Knopman, Ryan Keefe

Prepared for the United States Bureau of Reclamation

The research described in this report was prepared for the United States Bureau of Reclamation and conducted in the Environment, Energy, and Economic Development Program within RAND Justice, Infrastructure, and Environment.

**Library of Congress Cataloging-in-Publication Data** is available for this publication.

ISBN: 978-0-8330-8179-7

The RAND Corporation is a nonprofit institution that helps improve policy and decisionmaking through research and analysis. RAND's publications do not necessarily reflect the opinions of its research clients and sponsors.

**Support RAND**—make a tax-deductible charitable contribution at www.rand.org/giving/contribute.html

**RAND®** is a registered trademark.

*Cover image: Hoover Dam Water Electricity Power Station, © malajscy/Fotolia. The cover image shows the current low level of the Lake Mead reservoir and two of the four water intake towers that control the supply of water to the power plant turbines at the base of Hoover Dam. The line between the light rock and dark rock around the reservoir indicates the usually higher elevation of the water.*

© Copyright 2013 RAND Corporation

# Figures

S.1. Summary of Long-Term Water Delivery Outcomes That Do Not Meet Objectives ..... xvi

S.2. Trade-Offs Between Portfolio Costs and Vulnerabilities (2041–2060) Across Portfolios for the Upper and Lower Basins ................................ xviii

S.3. Percentage of Traces in Which Options Are Implemented and Associated Implementation Delay for *Portfolio D (Common Options)* ................... xx

1.1. Basin Study Area ................................................................. 2

1.2. Historical Supply and Use and Projected Future Colorado River Basin Water Supply and Demand ........................................................ 4

1.3. Steps to the Basin Study ........................................................ 6

2.1. Emerging Adaptive Decision Making Framework for Long-Term Water Planning and Management ................................................... 10

2.2. Steps in Robust Decision Making Process ................................... 11

3.1. Distribution of Annual Colorado River Streamflow (2012–2060) for Each Basin Study Supply Scenario ......................................... 18

3.2. Summary of Temperature Changes Across Traces (left) and Precipitation Changes Across Traces (right) for the Historical and Future Climate Scenarios ............ 19

3.3. Simplified Schematic of CRSS Network ..................................... 21

3.4. Example Simulations in Which Lake Mead Pool Elevation Objectives Are Met and Not Met .......................................................... 23

3.5. Prioritized List of Options for One Portfolio (*Portfolio A*) ................. 25

3.6. Schematic Representing Options Included in Each of the Four Portfolios ........ 26

3.7. Simulated Lake Mead Pool Elevation Over Time Without Options (red) and With Options Implemented by a Simple Dynamic Portfolio ..................... 29

4.1. Summary of Long-Term Water Delivery Outcomes That Do Not Meet Objectives ..... 32

4.2. *Declining Supply* Vulnerable Conditions for Lee Ferry Deficit (Streamflow Variables) ... 34

4.3. Vulnerable Conditions for Lee Ferry Deficit (Climate Variables) ............. 36

4.4. Low Historical Supply Vulnerable Conditions for Lake Mead Pool Elevation (Streamflow Variables) ......................................................... 38

4.5. Low Historical Supply Vulnerable Conditions for Lake Mead Pool Elevation (Climate Variables) ............................................................ 39

5.1. Reduction in Upper Basin Reliability Vulnerability (Lee Ferry Deficit) Across Portfolios ..................................................................... 44

5.2. Reduction in Lower Basin Vulnerability (Lake Mead Pool Elevations) Across Portfolios ..................................................................... 45

5.3. Change in Streamflow Conditions and Climate Conditions Leading to Upper Basin Vulnerability for *Portfolio A (Inclusive)* ................................ 46

5.4. Change in Streamflow Conditions and Climate Conditions that Lead to Lower Basin Vulnerability for *Portfolio A (Inclusive)* ........................... 47

5.5.    Percentage of Traces in Which Five Options Are Implemented by Year for *Portfolio D (Common Options)*................................................................................. 48

5.6.    Frequency of Option Implementation (percentage of traces) for *Portfolio D (Common Options)*................................................................................. 50

5.7.    Distribution of Total Annual Cost Resulting from Implementation of the Dynamic Portfolios for All Traces, Low Historical Supply, and Declining Supply ................... 51

5.8.    Trade-Offs Between Portfolio Costs and Vulnerabilities Across Portfolios ................. 52

6.1.    Percentage of Traces in Which Options Are Implemented and Associated Implementation Delay for *Portfolio D*.......................................................................57

A.1.    Example Simulations in Which Lake Mead Pool Elevation Objectives Are Met and Not Met................................................................................................................. 59

A.2.    20 Simulation Outcomes with Definitions of Vulnerable Conditions ...................... 61

A.3.    Classification of Ten Simulations by Vulnerable Conditions................................. 62

A.4.    Second Part of the Low Historical Supply Vulnerable Conditions for Lake Mead Pool Elevation Using Climate Variables ....................................................... 63

# Tables

S.1.   Summary of Uncertainties, Policy Levers, Relationships, and Metrics Addressed in Study (XLRM Matrix)................................................................. xiii

S.2.   Descriptions of Four Portfolios........................................................... xiv

3.1.   Summary of Uncertainties, Policy Levers, Relationships, and Metrics Addressed in Study (XLRM Matrix)................................................................. 15

3.2.   Demand Scenario Storylines, Scenarios, Descriptions, and Statistics...................... 16

3.3.   Supply Scenario Description and Number of Traces...................................... 17

3.4.   Experimental Design for Evaluating the Vulnerabilities of the Baseline Management Approach and with Portfolios.............................................. 20

3.5.   Performance Metrics and Vulnerability Thresholds for the Two Water-Deliveries Metrics Used in This Report................................................................ 22

3.6.   Descriptions of Four Portfolios........................................................... 25

3.7.   Summary of Options Included in the Four Portfolios.................................... 27

3.8.   Definitions of the Vulnerability Signposts Developed for Each Indicator Metric........ 28

4.1.   Vulnerable Conditions Defined for Lee Ferry Deficit: Declining Supply.................. 37

4.2.   Vulnerable Conditions Defined for Lake Mead Pool Elevation: Low Historical Supply..................................................................................... 40

5.1.   Option Types Included in the Portfolio Analysis....................................... 49

A.1.   Example Statistics for Several Different Example PRIM Vulnerable Conditions.......... 61

A.2.   Vulnerable Conditions Defined for Lake Mead Pool Elevation with No Continuation of Interim Guidelines........................................................ 64

B.1.   Description of Each Type of Option Included in the Portfolio Analysis.................. 65

# Summary

## Introduction

The Colorado River is the single most important source of water in the southwestern United States, providing water and power for nearly 40 million people. In recent decades, federal managers and Colorado River water users have grown increasingly concerned about the future reliability of the River's water supply. Demand for water in the Lower Basin (California, Arizona, and Nevada) already exceeds the 7.5 million acre-feet (maf) volume allocated in 1922 through the Colorado River Compact (the Compact)—the legal document that determines the allocation of water to the Upper Basin (Colorado, Utah, Wyoming, and New Mexico) and the Lower Basin. Demand also continues to grow in the Upper Basin states.

Water from the River was initially allocated based on two decades of unusually high river flow, meaning it is likely the River was significantly overallocated when the Compact was signed. In addition, an extended drought from 2000 to 2007 has reduced total water storage in Colorado Basin reservoirs from nearly full to 55 percent of capacity; the system remains just over half-full as of this writing. The combination of increasing demand and lower-than-expected streamflow has steadily eroded system resilience.

Moreover, a growing body of literature suggests the Colorado River system is now—or soon will be—operating in a new hydrologic regime for which past data and experience are not an adequate guide for future river conditions. Climate simulations applied in the Colorado River Basin Study (Basin Study) are generally consistent in indicating that the entire Basin will track global trends and become warmer, but climate simulations of regional precipitation changes in the Upper Basin—where most River source water falls as snow or rain—generate very different forecasts. Some models project precipitation *declines* of up to 15 percent over the next 50 years in the Upper Basin, while others forecast an *increase* in precipitation of up to 11 percent over that time. Despite this uncertainty, Basin shortages are projected to increase; the question remains how much and when.

Motivated by these challenges and in response to directives in the United States SECURE Water Act of 2009 (Public Law 111–11, 2009), the Bureau of Reclamation (Reclamation) and water-management agencies representing the seven Basin States initiated the Basin Study in January 2010 to evaluate the resiliency of the Colorado River system over the next 50 years (2012–2060) and compare different options for ensuring successful management of the River's resources.

However, in conducting this evaluation, Reclamation and the water agencies must deal not with a future that is uncertain but well understood; instead, they must plan for a future that is *deeply uncertain* and one that cannot be described statistically because of a lack of knowledge about how changes will unfold. Under these conditions, developing an optimal management strategy designed to perform well for a single deterministic or probabilistic forecast of future conditions is not very useful; rather, planners need a *robust* and *adaptive* strategy—robust in that it performs well over a wide range of possible futures and adaptive in that it can adjust over time in response to evolving conditions.

Given these circumstances, RAND was asked to joined the Basin Study Team in January 2012 to help develop an analytic approach to identify key vulnerabilities in managing the Colorado River Basin over the coming decades and to evaluate different options that could reduce these vulnerabilities. Building off the earlier Basin Study efforts, RAND applied an approach called Robust Decision Making (RDM)—a systematic, objective approach for developing management strategies that are more robust to uncertainty about the future. In particular, RAND researchers:

- identified future vulnerable conditions that could lead to imbalances that could cause the Basin to be unable to meet its water delivery objectives
- developed a computer-based tool to define "portfolios" of management options reflecting different strategies for reducing Basin imbalances
- helped evaluate these portfolios across a range of simulated future scenarios to determine how much they could improve Basin outcomes
- analyzed the results from the system simulations to identify key trade-offs among the portfolios.

This report summarizes RAND's contribution to the Basin Study (the *Colorado River Basin Water Supply and Demand Study* was released in December 2012). In contrast to Reclamation's report—which covers the entire Basin Study and comprises seven primary documents, dozens of appendixes, and thousands of pages of results—this document is intended to concisely summarize RAND's evaluation of long-term water delivery reliability for the Colorado River Basin across the range of future uncertainties and with proposed new options in place. This report focuses more on the analysis of vulnerabilities and how this information can inform the development of a robust management strategy for the Colorado River Basin. We worked closely with the Basin Study Team and state partners to complete this analysis. Here, we use only a small subset of the study results to tell the story of emerging water supply vulnerability and possible actions to reduce vulnerability. For example, although the Basin Study developed a wide range of performance metrics, we considered only broad, high-level performance metrics—each representing delivery reliability for the Upper and Lower Basins.

## Developing Robust Management Strategies for the Colorado River Basin

RDM uses a framework called XLRM to summarize scenarios developed to reflect future uncertainty (X), the options (L) evaluated that would compose a robust management strategy, the model used to simulate future conditions (R), and the performance metrics (M) used to evaluate system robustness. Table S.1 shows the XLRM framework for this effort; a much larger set of performance metrics were used in the full Basin Study, but here we focus on two of the key ones to simplify the discussion of RDM's contribution.

**Table S.1**
**Summary of Uncertainties, Policy Levers, Relationships, and Metrics Addressed in Study (XLRM Matrix)**

| Uncertainties or Scenario Factors (X) | Management Options and Strategies (L) |
|---|---|
| Demand for Colorado River water<br>Future streamflow or water-supply climate drivers<br>Reservoir operations post-2026 | Current Management<br>Four portfolios composed of individual options<br>• Demand reduction<br>• Supply augmentation |
| **Relationships or Systems Model (R)** | **Performance Metrics (M)** |
| Colorado River Simulation System (CRSS) | *Upper Basin Reliability*—Lee Ferry Deficit<br>*Lower Basin Reliability*—Lake Mead Pool Elevation<br>Cost of option implementation |

## Scenarios and Uncertainty (X)

During the first year of the study (and before RAND was involved), the Basin Study Team developed a set of supply, demand, and reservoir operations scenarios designed to capture the uncertainties planners face. Each scenario describes one plausible way that each of these three factors could evolve over the study's 49-year time horizon (2012–2060).

The Basin Study Team developed *four supply scenarios* based on different sources of future streamflow estimates. Each scenario is composed of many different 2012–2060 time series of streamflows—known as *future traces or traces*. The first scenario, Historical, is based on the *recent historical record*. Each trace within the Historical scenario is a repeat of the historical record (from 1906 to 2007) with a different starting year. The second and third scenarios are based on streamflow estimates derived from *paleoclimatological proxies*, such as tree ring data. Each trace is consistent with a subset of years from the paleoclimatological record. The fourth scenario is derived from the projections of *future climate conditions* from 16 global climate models and three global carbon emissions projections. Each trace is derived from downscaled results from a single general circulation model (GCM) projection and emissions scenario.

The Basin Study Team also developed *six demand scenarios* that span a range of plausible future demands, not considering additional programs and incentives for water conservation: (1) current projected growth; (2) slow growth with an emphasis on economic efficiency; (3) rapid growth due to economic resurgence; (4) rapid growth with current preferences toward human and environmental values; (5) enhanced environment due to expanded environmental awareness; and (6) enhanced environment due to stewardship with growing economy. As input to the vulnerability analysis, RAND calculated the average demand in the last two decades of each trace (2041–2060). The post-2040 demand ranges from 13.8 maf (slow growth) to 15.6 maf (rapid growth).

Lastly, *two reservoir operations scenarios* were created, reflecting different assumptions about how the system would be operated beyond 2026, when the 2007 Interim Guidelines are scheduled to expire. In one, the guidelines for Lower Basin shortage allocation and reservoir management are extended; in the other, they instead revert to the "No Action" Alternative as stipulated in the 2007 Interim Guidelines Environmental Impact Statement (EIS). Continuation of the Interim Guidelines means the continuation of mandatory, agreed-upon Lower Basin shortages to help maintain storage in Lake Mead if the lake elevation drops below 1,075 feet above mean sea level (msl).

When evaluating the performance of the Colorado River Basin system, the four supply scenarios, six demand scenarios, and two reservoir-operations scenarios were combined and totaled 23,508 individual traces.

### Options and Strategies to Improve Performance (L)

The Basin Study evaluated the baseline reliability of the Colorado River system by simulating current operating rules and procedures—what is referred to as the Current Management baseline (as shown in Table S.1). It also evaluated a wide array of different supply-augmentation and demand-reduction options that could improve system performance and reduce vulnerabilities. Such options were organized into eight categories: (1) agricultural conservation, (2) desalination, (3) energy water use and efficiency, (4) water imports into basin, (5) local supply, (6) municipal and industrial (M&I) conservation, (7) reuse, and (8) watershed management. Starting with 150 different options, the Basin Study Team ultimately evaluated a smaller set of these options—about 80—according to cost, yield, availability, and 16 other criteria, including technical feasibility, permitting risk, legal risk, policy risk, and energy intensity.

The RAND team developed a "Portfolio Development Tool" that was used by the Basin Study Team and stakeholders to develop four strategies defined by portfolios of prioritized supply-augmentation and demand-reduction options (drawn from the 80 evaluated ones): *Portfolio A (Inclusive), Portfolio B (Reliability Focus), Portfolio C (Environmental Performance Focus)*, and *Portfolio D (Common Options)* (Table S.2).

To evaluate how each portfolio of options would perform across the wide range of futures, the Basin Study Team defined *dynamic portfolios*, which include rules within the simulation model used in this study to implement options only when conditions indicate a need for them. The RAND and Study Team developed a set of "signposts" for six different water delivery metrics, including the two discussed in this report—Lee Ferry Deficit and Lake Mead Pool Elevation. Signposts specify a set of observable system conditions and thresholds that indicate that vulnerabilities are developing. During a simulation, the model monitors the signpost conditions; if any thresholds are crossed, then it implements options from the top of the portfolio option list. In this way, the dynamic portfolios seek to more realistically mimic how options would be implemented over time in response to system needs.

**Table S.2**
**Descriptions of Four Portfolios**

| Portfolio Name | Portfolio Description |
|---|---|
| *Portfolio A (Inclusive)* | Includes all options included in the other portfolios |
| *Portfolio B (Reliability Focus)* | Emphasizes options with high technical feasibility and high long-term reliability; excludes options with high permitting, legal, or policy risks |
| *Portfolio C (Environmental Performance Focus)* | Excludes options with relatively high energy intensity; includes options that result in increased instream flows; excludes options that have low feasibility or high permitting risk |
| *Portfolio D (Common Options)* | Includes only those options common to Portfolio B (Reliability Focus) and Portfolio C (Environmental Performance Focus). |

NOTE: The portfolio names in parentheses were developed for this report only. The *Colorado River Basin Water Supply and Demand Study* used only the lettered names (Reclamation, 2012f, 2012h).

### Simulating the Colorado River System and Performance Metrics (R and M)

The Basin Study used the Colorado River Simulation System (CRSS), Reclamation's long-term planning model, to simulate the Colorado River system. CRSS estimated the future performance of the system with respect to a large set of different types of performance metrics—*water deliveries* (nine metrics), *electric power resources* (two metrics in three locations), *water quality* (one metric in 20 locations), *flood control* (three metrics in ten locations), *recreational resources* (two metrics in 13 locations), and *ecological resources* (five metrics in 34 locations).

While the full Reclamation report used all the performance metrics, this report focuses on two key water delivery metrics—Lee Ferry Deficit and Lake Mead Pool Elevation. These were the metrics used in the Basin Study to compare the performance of options and strategies, as they broadly summarize the reliability of the Upper and Lower Basins, respectively. If there is a Lee Ferry deficit, then there could be delivery reductions in the Upper Basin to augment flows to the Lower Basin. The health of the Lower Basin system and deliveries to the Lower Basin states are similarly closely tied to the Lake Mead elevation.

## Future Vulnerabilities to Colorado Basin Water Deliveries

Using the RDM approach and inputs described above, RAND and the Study Team first evaluated the vulnerabilities of the Colorado River system. We addressed two key questions: (1) under which futures does the Basin not meet water delivery objectives, and (2) what future external conditions lead to vulnerabilities? Again, here we focus on the two key water delivery performance metrics.

### Under Which Futures Does the Basin Not Meet Water Delivery Objectives?

Figure S.1 summarizes Upper Basin Reliability (Lee Ferry Deficit) and Lower Basin Reliability (Lake Mead Pool Elevation) across all 23,508 traces representing future uncertainty in two ways: (1) the percentage of traces in which management objectives are not met at least once during the time period (left side), and (2) the percentage of all years in the simulation in which outcomes did not meet objectives (right side). For Upper Basin Reliability, the percentage of traces in which at least one Lee Ferry deficit occurs increases from 2 percent (from 2012 through 2026) to 16 percent (from 2041 through 2060), with Lee Ferry deficits occurring in 6 percent of the years (three years) in the last period (top half of the figure). Similarly, for Lower Basin Reliability, Lake Mead elevations fall below the 1,000-foot elevation threshold more frequently across traces and years in later periods.

### What Future External Conditions Lead to Vulnerabilities?

While the above analysis tells us how vulnerable the Current Management approach is over time, it does not tell us what external conditions lead to those projected vulnerabilities. Using RDM vulnerability analysis techniques and statistical summaries of streamflow at Lee Ferry, we looked for a set of future conditions that best captures the vulnerable traces. We find that the Upper Basin is susceptible to a Lee Ferry Deficit when two future conditions are met: long-term average streamflow declines beyond what has been observed in the recent historical record (below 13.8 maf per year) and there is an eight-year period of consecutive drought years where the average flow dips below 11.2 maf per year. Traces that meet both of these

**Figure S.1**
**Summary of Long-Term Water Delivery Outcomes That Do Not Meet Objectives**

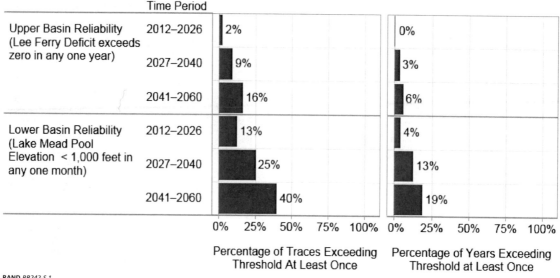

RAND *RR242-S.1*

conditions—called Declining Supply vulnerable conditions—lead to a Lee Ferry deficit 87 percent of the time.

Using the same approach, we find that Lake Mead elevation is vulnerable to conditions in which supplies are simply below the long-term historical average—specifically, when long-term average streamflow at Lees Ferry falls below 15 maf, and an eight-year drought with average flows below 13 maf occurs.[1] We call these conditions Low Historical Supply vulnerable conditions, and they describe 86 percent of all traces that lead to unacceptable results. We also defined vulnerable conditions for both the Upper Basin and Lower Basin delivery reliability using climate inputs to describe supply in the Historical and Future Climate supply scenarios.

## Reducing Vulnerabilities Through New Management Options

RAND and the Basin Study Team evaluated the four portfolios of supply-augmentation and demand-reduction options—*Portfolio A (Inclusive)*, *Portfolio B (Reliability Focus)*, *Portfolio C (Environmental Performance Focus)*, and *Portfolio D (Common Options)*—across all the scenarios described above. We next reviewed how each performed under the vulnerable conditions—Declining Supply and Low Historical Supply. We find that implementation of the portfolios reduces the number of years in which the system fails to meet Basin goals across many, but not all, scenarios.

---

[1]   Lee Ferry is close to, though slightly downstream from, the U.S. Geological Survey flow gauge at Lees Ferry, Arizona. The Paria River enters the Colorado River between these locations, leading to small differences in flows between the two points. In this report, we use "Lee Ferry" when referring to the Compact delivery requirements from the Upper to Lower Basin, and "Lees Ferry" when referring to natural streamflow measurements of the Colorado River.

## How Well Do Portfolios of Options Reduce Vulnerabilities?

For the Upper Basin Reliability metric—Lee Ferry Deficit—implementation of the portfolios reduces the percentage of years and traces in which deficits occur. *Portfolio C (Environmental Performance Focus)* is more effective than *Portfolio B (Reliability Focus)* in reducing vulnerabilities. For the Lower Basin Reliability metric—Lake Mead Pool Elevation—implementing the portfolios significantly reduces the number of years in which the Basin goals are not met. Even in the most stressing Declining Supply vulnerable conditions, the percentage of years is reduced from 50 percent to around 25 percent. These reductions in yearly vulnerability, however, do not lead to significantly fewer traces in which Lake Mead elevation drops below 1,000 feet in at least one year. The results also show that *Portfolio B (Reliability Focus)* is somewhat more effective at reducing Lower Basin vulnerability than *Portfolio C (Environmental Performance Focus)*.

The implementation of portfolios increases the robustness of the system and shrinks the set of conditions in which the system would not meet its goals. The Basin becomes less vulnerable to lower flow sequences and drying periods. In terms of climate conditions, with a portfolio in place, the Basin performs well over warmer and dryer climate conditions. Chapter Five provides more specific detail.

## What Are the Key Trade-Offs Among Portfolios?

How effective the portfolios are in reducing vulnerabilities is not the only criterion for assessing them. Implementation costs, which increase over time as options are implemented in response to the signposts, are another assessment criterion. There is a wide range in costs across the traces. For *Portfolio A (Inclusive)*, for example, the costs range from just under $2 billion per year to more than $7 billion per year in 2060. This wide range of costs indicates that the dynamic portfolios as designed for the study help restrain unnecessary investment in futures when conditions do not warrant it.

One of the advantages of the RDM approach is that it allows us to combine the cost and vulnerability results together to draw out the distinctions and trade-offs among the four portfolios. Figure S.2 shows total annual implementation costs in 2060 for the four portfolios (the horizontal axis) and the percentage of years vulnerable from 2041 to 2060 (the vertical axis) for all traces and for the two vulnerable conditions. We are looking for portfolios that have the lowest costs (farthest to the left in all the graphs) and that reduce vulnerabilities the most (the lowest on all the graphs). The portfolios are distinguished by color here, with the labeling shown in the bottom band in the figure.

As shown in the figure, we find little difference among portfolios when looking across *all traces* evaluated. That is, the range in vulnerability reduction and costs overlap significantly for all the portfolios (the top band in the figure). This is not surprising because there are many traces evaluated in which there is only a modest need for improvement. All four of the portfolios can address those needs using options with similar costs.

However, when we focus on traces corresponding to the two vulnerable conditions, we see some differences across the portfolios. First, in the Low Historical Supply conditions (the middle band in the figure), we see that the portfolio with the most options (*Portfolio A*) most reduces the number of years in which the Upper Basin and Lower Basin goals go unmet. The ranges in costs (horizontal spread) across the traces increase significantly, but there is again significant overlap among the portfolios.

**Figure S.2**
**Trade-Offs Between Portfolio Costs and Vulnerabilities (2041–2060) Across Portfolios for the Upper and Lower Basins**

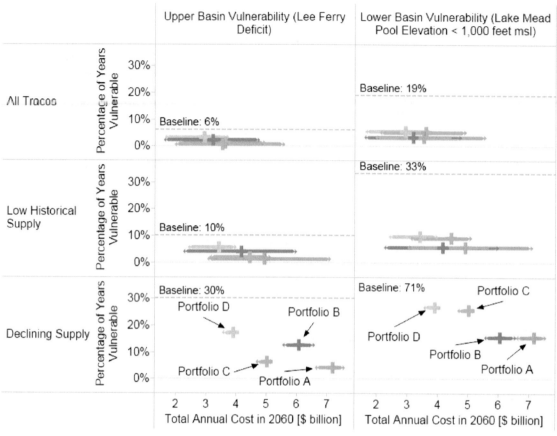

RAND *RR242-S.2*

When we only include traces in the *Declining Supply* vulnerable conditions (the bottom band in the figure), the trade-offs become clear. For the Upper Basin (left panel of the figure), *Portfolio C (Environmental Performance Focus)* is not only more effective than *Portfolio B (Reliability Focus)* and *Portfolio D (Common Options)*, it costs significantly less than *Portfolio B (Reliability Focus)*. Only *Portfolio A (Inclusive)* reduces vulnerability more, but it does so at significantly higher cost. *Portfolio C (Environmental Performance Focus)* dominates because it includes an Upper Basin water bank, which is used at Lee Ferry to maintain flow to the Lower Basin and excludes other, more expensive, new supply options (discussed more in Chapter Six).

However, performance with respect to the Lower Basin objectives in the Declining Supply vulnerable conditions (the bottom band in the figure, right panel) shows that *Portfolio B (Reliability Focus)* improves reliability as well as or better than the other portfolios in all three sets of conditions. *Portfolio B (Reliability Focus)* includes more options that directly benefit the Lower Basin, including Pacific Ocean desalination projects. Given this more focused investment, *Portfolio B (Reliability Focus)* dominates *Portfolio A (Inclusive)* by being just as effective but less costly.

## Implementing a Robust, Adaptive Strategy for the Colorado River Basin

The CRSS simulations of portfolios reveal traces in which options are implemented. Options that are implemented across many traces soon after they become available can provide the foundation of an initial robust strategy. We focus this analysis on the two vulnerable conditions (i.e., Declining Supplies and Low Historical Supplies) identified by this study, because these represent conditions when options are generally needed to alleviate system imbalances.

### Identification of Near-Term Options as a Foundation of a Robust Strategy

For each portfolio, we identified those options that are almost always needed regardless of differing assumptions about future conditions. Because *Portfolio D (Common Options)* includes only options selected for both of the two stakeholder-derived portfolios (*Portfolios B and C*), options always or frequently implemented in this portfolio as soon as they are available can be considered both near-term and high priority.

Figure S.3 summarizes how frequently options from *Portfolio D (Common Options)* are implemented by 2060 (horizontal axis) and the delay in their implementation (vertical axis), expressed as the median delay across all traces relative to the time they become available. The results are presented for three sets of traces—all traces (top panel), those traces in the Low Historical Supply vulnerable conditions (middle panel), and those traces in the Declining Supply vulnerable conditions (bottom panel).

Results in the lower-right corner of the all traces panel (bounded by five years or less and 75 percent implemented or more) are near-term, high-priority options. In this case, M&I Conservation is shown to be required in more than 90 percent of all traces examined in the study with a minimum delay of only one year. Agricultural Conservation with Transfers is implemented in almost 100 percent of traces, but with a delay of six years. Three desalination options—Desal–Salton Sea, Desal-Yuma, and Desal-Groundwater—are all high-priority but are needed only after delays of eight years or more.

For future conditions consistent with the two key vulnerable conditions—Low Historical Supply and Declining Supply—more options are needed, with less delay. The middle panel of Figure S.3 shows that for the Low Historical Supply vulnerable conditions, the urgency of implementation of Agricultural Conservation with Transfers and Desal-Salton Sea increases, making them both near-term, high-priority options. The Reuse-Municipal option is also required in more than 70 percent of traces. The bottom panel shows that for Declining Supply vulnerable conditions, all options in *Portfolio D (Common Options)* are needed by 2060 in nearly all traces.

Figure S.3 shows that most of the options in the *Portfolio D (Common Options)* are needed in only some future traces and in many cases are implemented only after a delay. However, the conditions corresponding to the Low Historical Supply vulnerable conditions have been experienced in the recent past and those corresponding to the Declining Supply are predicted by many global climate model simulations. As the Basin Study highlights, the Basin does not need to commit to all possible options now, but it might use the available lead time to prepare to invest in new options if conditions suggest they are warranted. The implementation of some options with longer lead times will need to be initiated soon so they would be available if needed under particular future traces. Exploring plans during this time for design and permitting of selected options would provide decision makers with a hedge against potential delays in implementation if the options are needed in response to changing conditions.

**Figure S.3**
**Percentage of Traces in Which Options Are Implemented and Associated Implementation Delay for *Portfolio D (Common Options)***

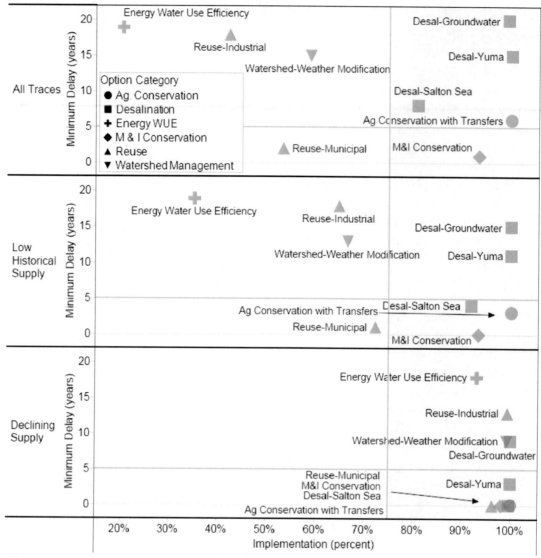

NOTE: Ag = agricultural; Desal = desalination; WUE = water use efficiency.

RAND *RR242-S.3*

## Monitoring Conditions to Signal Implementation of Additional Options

Reclamation and other agencies are already collecting critical information (e.g., streamflow, climate conditions, status of the reservoirs) that can be used to inform assessments of which options should be implemented in the future. Building this information into systematic and recurring system assessments would enable managers and users of the Basin to better understand how conditions are evolving and plan for additional management options accordingly.

The vulnerability analysis specifically showed that the Upper Basin is vulnerable to climate conditions that are consistent with many of the simulated conditions emerging from a variety of global climate models. Over the next few years, new climate models or higher-resolution regional climate projections might make it easier to discern whether the future climate is going continue to deviate from the historical record. If the results from improved models are consistent with the more pessimistic current projections, the Basin is increasingly likely to face vulnerable conditions for the Lee Ferry Deficit and Lake Mead Pool Elevation levels. Many of the options identified as necessary under these conditions would need to be considered for implementation.

### Options to Implement If Future Conditions Warrant

The analysis has shown that as vulnerable conditions develop in the Basin, increasingly expensive adaptations will be required. The analysis highlighted which options would be needed and when. However, for many of these options, preparation would need to begin well before the time of implementation. For this mid- to longer-term implementation period of a robust, adaptive strategy, Reclamation and the Basin States could identify the key long lead-time options that may be needed and begin to take near-term planning and design steps to ensure their availability.

It may also be beneficial to consider additional management and governance-based approaches for addressing future imbalances. Many of these options, such as some types of water transfers, could be consistent with the current Law of the River, but could not be easily modeled by CRSS within the time available to complete the study. As suggested by the Basin Study, evaluating these additional options in the coming months could further improve the ability for the portfolios to address supply and demand imbalances. Revisiting the options included in the portfolio is fully consistent with the RDM analysis framework used in the Basin Study. Comparing and contrasting the performance and other attributes of additional approaches alongside the adaptive options evaluated for the Basin Study would support the successful implementation of a robust, adaptive strategy.

# Acknowledgments

The authors are grateful to the Bureau of Reclamation's Terry Fulp, Director, Lower Colorado Region, for his support of RAND's involvement in the Basin Study, and to Carly Jerla, Colorado River Basin Study Director, for her leadership of the Colorado River Basin Study. This work represents significant collaboration with the other members of the study team, Jim Prairie (Bureau of Reclamation), Ken Nowak (Bureau of Reclamation), Alan Butler (Bureau of Reclamation), Armin Munevar (CH2MHill), Klint Reedy (Black and Veatch), and the Center for Advanced Decision Support for Water and Environmental Systems (CADSWES). Pamela Adams and Kaylee Nelson from the Bureau of Reclamation also provided critical support through the project. The Basin Study state representatives also provided invaluable guidance and feedback. Finally, we would like to thank our reviewers, Dr. Nick Burger of RAND and Dr. Julien Harou at the University College London, for their constructive comments; Paul Steinberg for his assistance developing this manuscript; and Keith Crane, Director of the Environment, Energy, and Economic Development Program, for his guidance throughout the effort and review of this report.

# Abbreviations

| | |
|---|---|
| af | acre-feet |
| ag | agricultural |
| Basin Study | Colorado River Basin Study |
| Compact | Colorado River Compact |
| CRBS | Colorado River Basin Study |
| CRSS | Colorado River Simulation System |
| EIS | Environmental Impact Statement |
| GCM | General Circulation Model |
| ICS | intentionally created surplus |
| IQR | interquartile range |
| ISM | Indexed Sequential Method |
| LB | Lower Basin |
| M&I | municipal and industrial |
| maf | million acre-feet |
| msl | mean sea level |
| PRIM | Patient Rule Induction Method |
| RDM | Robust Decision Making |
| Reclamation | U.S. Bureau of Reclamation |
| taf | thousand acre-feet |
| UB | Upper Basin |
| WUE | water use efficiency |
| XLRM | matrix of uncertainty factors (X), decision options (levers) and strategies (L), relationships (R), and performance metrics (M) |

# Glossary of Selected Terms

| | |
|---|---|
| baseline management | the current Colorado River management system that serves as a baseline for the assessment of system vulnerabilities |
| case | a single simulation of the water-management model over the 49-year study period |
| future | uncertain future conditions that arise in response to specified scenarios (e.g., supply, demand, and reservoir operations scenarios) |
| option | a specific investment or program to increase Basin supply, reduce Basin demand, or affect operations |
| performance metric | a quantitative variable that indicates the functioning of the water-management system |
| portfolio | a specific set of water-management options to be implemented over time by the water-management model in response to emerging vulnerabilities |
| scenario | a description of uncertain future parameters pertaining to, for example, supply, demand, and reservoir operations |
| strategy | a specific approach, in terms of the types of options used, for addressing supply and demand imbalances |
| trace | a single time series of climate or hydrological variables that is used to simulate the time evolution of the water-management system |
| vulnerability | an outcome that is not consistent with the Basin's management goals |
| vulnerable conditions | those uncertain future conditions that generally lead to vulnerabilities under baseline management |

# CHAPTER ONE

# Introduction

## Managing the Colorado River Basin—A History of Reliability

The Colorado River is the single most important source of water in the southwestern United States, providing water and power for nearly 40 million people and water to irrigate more than five million acres of farmland across seven states and for 22 Native American tribes (National Research Council, 2007; U.S. Bureau of Reclamation [Reclamation], 2012h). The River supports billions of dollars of economic activity—irrigating 15 percent of U.S. crops, for example—and is also the lifeline for two dozen National Parks, Wildlife Refuges, and Recreation Areas (Reclamation, 2012h).

The Colorado River system is made up of the River itself, tributary streams and rivers, and water storage and delivery infrastructure (dams and reservoirs, hydropower facilities, canals, aqueducts, and pumps). Significant infrastructure on the River includes Lake Powell in Utah (Glen Canyon Dam), Lake Mead in Nevada (Hoover Dam), the Central Arizona Project (which delivers water from the River to Arizona farms and municipalities), and the Colorado River Aqueduct and All-American Canal, which collectively divert water to Southern California users (Figure 1.1). Much of this infrastructure is operated and maintained by the U.S. Bureau of Reclamation (Reclamation), the agency that helps manage the Colorado Basin system and ensure that major water users reliably receive their water deliveries each year.

Water from the Colorado is apportioned to users in the seven Colorado River Basin States and adjacent areas that receive river water according to a series of federal laws and agreements, beginning with the Colorado River Compact of 1922 (the Compact). Based on two decades of unusually high river flow, Compact negotiators believed the natural flow of the Colorado River to be about 16.4 million acre-feet (maf) per year on average in 1922 (MacDonnell, Getches, and Hugenberg, 1995).[1] Using this estimate, the Compact initially allocated 15 maf of water equally among Colorado, New Mexico, Utah, and Wyoming (known as the Upper Basin States), and Arizona, California, and Nevada (known as the Lower Basin States). Each basin is entitled to consumptive use of 7.5 maf per year (Figure 1.1).[2]

In the first decade after the Compact was signed, water was used primarily for agricultural development, and specific allocations of water within the states were still loosely defined.

---

[1] An acre-foot (af) is the volume of water that covers one acre to a depth of a foot: 43,560 cubic feet, 325,853 U.S. gallons, or 1,233 cubic meters. The flow of the Colorado River is often measured at a gauge operated by the U.S. Geological Survey at Lee Ferry, Arizona, just downstream of Lake Powell. Approximately 92 percent of runoff enters the River upstream of this flow gauge, which is located near the dividing line between the Upper and Lower Basins (Reclamation, 2012b).

[2] Consumptive use is defined as the beneficial use of water that does not flow back or otherwise return to the River for downstream use.

**Figure 1.1**
**Basin Study Area**

SOURCE: Reclamation, 2012h.
RAND RR242-1.1

The Compact evolved over subsequent decades through additional federal acts, treaties, contracts, court decisions, and agreements—collectively referred to as the "Law of the River"—to accommodate the increasing demands placed on the system by urban development and natural streamflow variability. In 1944, the United States signed a treaty guaranteeing delivery of 1.5 maf per year to meet Mexico's water needs (United States, Mexico, 1944).

More recently it has become clear that the level of flow assumed for allocating water from the River—16.4 maf per year —is significantly higher than the actual long-term average flow. First, flow measurements over the hundred-year period of record from 1906 to 2005 records show an average flow of about 15 maf per year. Furthermore, reconstructions of river flow from the paleoclimatological record (time periods prior to direct measurement of flow), developed in the last 30 years based on tree-ring data, suggest that natural flow over the span of more than a thousand years was lower than recent observations: 13.5 to 14.7 maf per year (Stockton and Jacoby, 1976; Woodhouse, 2003; Meko et al., 2007; Woodhouse, Gray, and Meko, 2006). Together, these revised "streamflow" estimates suggest that the River was significantly overallocated when the Compact was signed.

Despite the gap between the historical average annual water supply and the initial Compact allocations, there has never been a large-scale water delivery shortage for Lower Basin water users. This historical reliability results from two factors. First, although the Lower Basin states have used at least their full allocation of water for decades, the Upper Basin states have never used their full apportionment of 7.5 maf. Actual consumptive use in the Upper Basin has grown over time, but only reached 3.8 maf in 2010, not including evaporation losses (Reclamation, 2012c). Because the Secretary of the Interior has the authority to redistribute allocated water that goes unused in any given year, water not used in the Upper Basin has been available for redistribution to Lower Basin states such as California and Nevada that have in the past requested more water than legally allocated by the Law of the River (Glennon and Pearce, 2007). Second, the Basin has approximately 65 maf of storage capacity in Lakes Powell and Mead and other reservoirs. When full, these storage reservoirs could provide water to meet roughly four years of demand even without new streamflow. Over the last 90 years, water from these reservoirs has provided a sufficient buffer to prevent shortages even during extended drought periods.

## Near- and Long-Term Threats to the Sustainable Use of the River

In recent decades, federal managers and Colorado River water users have grown increasingly concerned about the future reliability of the water supply. Demand for Colorado River water in the Lower Basin exceeds its 7.5 maf water allocation in the Compact, and demand continues to grow in the Upper Basin states (Reclamation, 2012c). The combination of increasing demand and lower-than-expected streamflow has steadily eroded the robustness of the system, as shown by historical supply and use through 2008 (left half of Figure 1.2). Critically, an extended drought from 2000 to the present has reduced total water storage in Colorado Basin reservoirs from nearly full to about 50 percent of capacity as of this writing.[3]

---

[3]  Reclamation estimates that by October 1, 2013, the total system storage will be at 49 percent of capacity (28.8 maf) (Reclamation, 2013).

**Figure 1.2**
**Historical Supply and Use and Projected Future Colorado River Basin Water Supply and Demand**

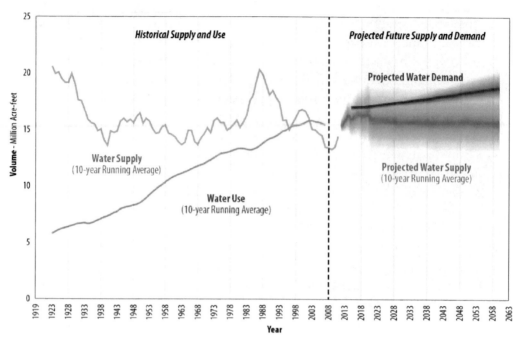

SOURCE: Reclamation, 2012h.
RAND *RR242-1.2*

Recognizing the growing challenge, Reclamation and the seven Basin States agreed in December 2007 to a revised, temporary set of management guidelines for the Basin through the year 2026, entitled the *Colorado River Interim Guidelines for Lower Basin Shortages and the Coordinated Operations for Lake Powell and Lake Mead* (Interim Guidelines) (Reclamation, 2007). The Interim Guidelines specified a new mechanism for sharing shortages among Lower Basin users in years when there is low reservoir storage in the system. The Interim Guidelines also revised the rules for balancing water volumes in Lakes Powell and Mead and began a new system of intentionally created surplus (ICS). ICS is designed to allow large Lower Basin users to augment system water with new sources, offsets, or intentional reductions in deliveries that are instead banked in the system for future use.

Though an important first step toward a more flexible and resilient system, the Interim Guidelines have not eased concerns about the long-term reliability of Colorado River water deliveries. In addition to growing demand and historical overallocation, climate change threatens to alter the characteristics of basin hydrology and to reduce long-term average river flows, further threatening sustainable river use.

A growing body of literature suggests that the Colorado River system is now—or soon will be—operating in a new hydrologic regime for which past data and experience are not an adequate guide for future river conditions (Milly et al., 2008). Climate simulations applied in the Basin Study are generally consistent in indicating that the entire Basin will track global trends and become warmer, with increases of 1.0 to 1.5 degrees Celsius by 2025 and 2.5 to 4.0 degrees Celsius by 2080 relative to the 1971–2000 historical average (Reclamation, 2012b). A warming climate is expected to lead to reduced snowpack, a shift in the timing of snowmelt and runoff earlier in the year, and an overall increase in demand across a variety of water uses

(Reclamation, 2012b, 2012c). Recent studies have also suggested that climate change will lead to persistent drying in the Southwest over the next century as temperatures rise (Christensen and Lettenmaier, 2007; Christensen et al., 2004; Nash and Gleick, 1993; Seager et al., 2007) and that climate change may have already begun to affect important hydrological characteristics of the system (Barnett et al., 2008; Hoerling and Eischeid, 2006). Although these studies vary in their estimates, most suggest an average streamflow reduction of 10 to 20 percent over the next 50 to 100 years in response to a temperature increase between 1 and 4 degrees Celsius.

The climate simulations of regional precipitation changes in the Upper Basin—where most Colorado River source water falls as snow or rain—are not yet sufficiently detailed or accurate to confirm a drying trend over time. In general, average precipitation in the Upper Basin trends downward when projected across a range of simulations based on downscaled General Circulation Model (GCM) output. However, these simulation results remain highly divergent. Some models show precipitation declines of up to 15 percent over the next 50 years in the Upper Basin, while others show increases in precipitation of up to 11 percent over that time (Reclamation, 2012b). Reduced precipitation would exacerbate the management challenges posed by increasing demand and warmer temperatures. Alternatively, given the large amount of storage available in Lakes Powell and Mead, and elsewhere in the Basin, a wetter Upper Basin could help ease management challenges and improve delivery reliability from the Upper Basin to Lower Basin at Lee Ferry, or to Lower Basin users from Lake Mead.

Given this range of possible outcomes, substantial uncertainty and disagreement remain about the effects of climate change on near- and long-term basin management decisions and the best path forward to ensure reliable and sustainable future use (Barnett and Pierce, 2009, 2008; Rajagopalan et al., 2009; Wildman and Forde, 2012).

The second part of Figure 1.2 captures the substantial uncertainty about projected water demand and supply out to the year 2063. As the figure shows, uncertainty in supply (the wider blue band) is much greater than uncertainty in demand (the narrower red band).

## Colorado River Basin Study

Motivated by the challenges described above and in response to directives in the United States SECURE Water Act of 2009 (Public Law 111–11, 2009), Reclamation and water-management agencies representing the seven Basin States initiated a study in January 2010 to evaluate the resilience of the Colorado River system over the next 50 years and compare different options and strategies for ensuring successful management of the River's resources. Specifically, the *Colorado River Basin Water Supply and Demand Study* (Reclamation, 2012h) took the following broad steps:

- defined a range of possible outcomes for future basin water supply, demand, and other key conditions over the next 50 years
- evaluated the resilience of the Colorado River system over this period in terms of a broad range of system performance metrics for basin water deliveries, power generation, recreation, ecosystems, and other outcomes.
- identified and quantified key vulnerabilities that could threaten future basin management
- assembled information about potential options for addressing potential gaps between supply and demand

- evaluated how different combinations of proposed options could improve future outcomes and better ensure successful management of the River's resources.

The progression of the investigation is summarized in Figure 1.3 below, starting with the framing of the study and moving through the development of water demand and supply scenarios, the identification of system reliability metrics, and the identification, characterization, and evaluation of options and portfolios of options. The steps are described in a series of technical reports published by Reclamation (shown in the figure as Plan of Study and Technical Reports A–G), with the overall study summarized in the Study Report represented at the bottom of the flowchart (Reclamation, 2012h, 2012a–g).

The Basin Study represents Reclamation and the Basin States' first attempt to systematically evaluate and explore the wide range of possible outcomes for the Colorado River given the current level of uncertainty about the future. Although no management decisions will be directly made based on this study, it is intended to lay the groundwork for future decision making by Reclamation and River users.

**Figure 1.3**
**Steps to the Basin Study**

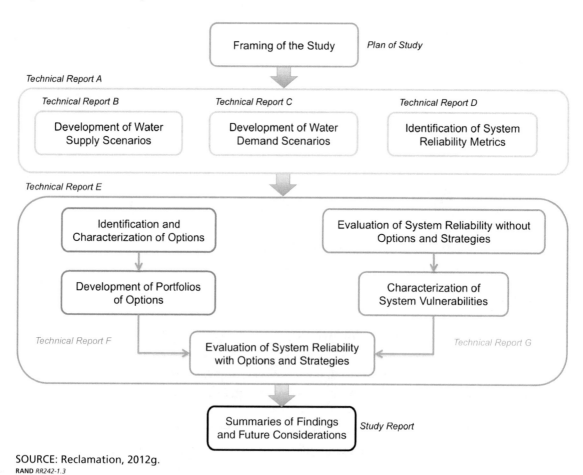

SOURCE: Reclamation, 2012g.
RAND RR242-1.3

## RAND's Role in the Basin Study

This effort is the first time that Reclamation has employed scenario analysis to grapple with future uncertainty at this scale, yielding a large number of potential scenarios to consider and options to evaluate. To meet this challenge, Reclamation and the Basin States asked researchers from the RAND Corporation to join the Basin Study Team (made up of Reclamation and supporting contractors) and apply quantitative scenario planning methods not yet applied in Reclamation studies to help understand and plan for the wide range of possible conditions currently projected for the Colorado River.

The approach that RAND implemented is called Robust Decision Making (or RDM), and is discussed in more detail in Chapter Two. RDM helps water managers iteratively identify and evaluate *robust* strategies—those that perform well in terms of management objectives over a wide range of plausible futures but may perform less well under an assumption that one future may be most likely to occur. This approach is ideally suited to address the significant and deep uncertainty that Reclamation and the Basin States face when planning for the next 50 years.

RAND joined the Study Basin Team in January 2012; roughly two years after the study began. By that point, the scope of the study was already established, and the Basin Study Team, together with state partners and other participating organizations, had identified and quantified key scenarios related to future water supply, demand, and reservoir management. RAND began by structuring an approach to navigate the large volume of information being produced to meet the study's overall goals, and focused the analysis on identifying key drivers that could lead to future vulnerabilities and the ability of proposed options to reduce these vulnerabilities. Specifically, RAND researchers:

- identified future conditions that could lead to the Basin not meeting its water delivery objectives
- developed a computer-based tool to define "portfolios" of management options reflecting different strategies for reducing Basin imbalances
- helped evaluate these portfolios across the range of scenarios to determine how much they could improve Basin outcomes
- analyzed the results from the simulations of the system to identify key trade-offs among the portfolios.

This report summarizes RAND's contribution to the Basin Study. In contrast to the study itself—which covers the entire Basin Study and comprises seven primary documents, dozens of appendixes, and thousands of pages of results—this document is intended to concisely summarize RAND's evaluation of long-term water delivery reliability for the Colorado River Basin across the range of future uncertainties and with proposed new options in place. This report focuses more than the Basin Study on the analysis of vulnerabilities and how this information can inform the development of a robust management strategy for the Colorado River Basin. We worked closely with the Basin Study Team and state partners to complete this analysis. We use only a small subset of the study results here to tell the story of emerging water supply vulnerability and possible actions to reduce vulnerability. For example, although the Basin Study developed a wide range of performance metrics, we considered only broad, high-level performance metrics—each representing delivery reliability for the Upper and Lower Basins.

To guide this investigation, we considered the following research questions:

- What future conditions threaten the reliability of Colorado River Basin water deliveries?
- How well do portfolios composed of diverse options reduce these vulnerabilities?
  - What are the key trade-offs in terms of cost and other option attributes?
  - What vulnerabilities still remain?
- What near-term actions would provide the foundation for a more robust strategy?
- What external conditions could be monitored to guide the implementation of the robust strategy?

## Report Organization

This report is divided into six chapters. Chapter Two describes the Robust Decision Making methods we applied to test future water deliveries against many scenarios and with different options in place. Chapter Three describes the scope of this portfolio of the Basin Study, including the uncertainties considered, options tested, and performance metrics evaluated. In Chapter Four, we describe the conditions that would most often lead to Reclamation and the Basin States not meeting their water delivery objectives. Chapter Five then shows how different portfolios of options could improve outcomes when faced with some of these stressing conditions and highlights key trade-offs between different approaches. Finally, in Chapter Six we discuss how the information developed for the Basin Study can help inform the implementation of a robust, adaptive strategy for the Colorado River.

# CHAPTER TWO

# Long-Term Water Planning and Management Under Uncertainty

In this section, we discuss the overarching approach used to address the deep uncertainty surrounding long-term water planning and management that Reclamation and the Basin States face. This provides the foundation for the discussion in Chapter Three about the specific approach used to develop robust management strategies for the Colorado River Basin.

## Why Robust and Adaptive Strategies Are Necessary

Although the flows on the Colorado River can vary substantially from year to year or decade by decade, conditions over the coming century could change in ways that differ significantly from what the region has experienced over the last hundred years. In addition, there are new elements of uncertainty about the Colorado River's future outlook driven by the confluence of increasing demand for water over time and a changing hydrology due to climate change.

This level of uncertainty has implications for the kind of long-term planning needed. Specifically, as managers and users of the Colorado River plan over the coming decades, they need to move beyond the traditional prediction-based approaches that work well when the future is uncertain but well understood. Instead, they must plan for a future that is deeply uncertain—one that will change in ways that cannot be described statistically because we do not know enough about how changes will unfold (Lempert, Popper, and Bankes, 2003). Rather than developing an optimal management strategy designed to perform well for a single deterministic or probabilistic forecast of future conditions, planners need a *robust* and *adaptive* strategy—robust in that it performs well over a wide range of possible futures, and adaptive in that it can adjust over time in response to evolving conditions. The Basin Study defines a strategy as a specific approach for augmenting supply or reducing demand through the implementation of different water-management options.

Figure 2.1, for example, illustrates the steps of an emerging framework for addressing climate change in long-term natural resource plans (National Academies of Science, 2011). It describes a series of iterative steps in which risks and options are evaluated; near-term decisions are made and implemented; and conditions are monitored to help refine those plans over time. This framework recognizes the importance of iterating, both in making management decisions (Steps 1 to 6, and back to 1) and in implementing successful strategies (Steps 6 to 8, and back to 1).

**Figure 2.1**
**Emerging Adaptive Decision Making Framework for Long-Term Water Planning and Management**

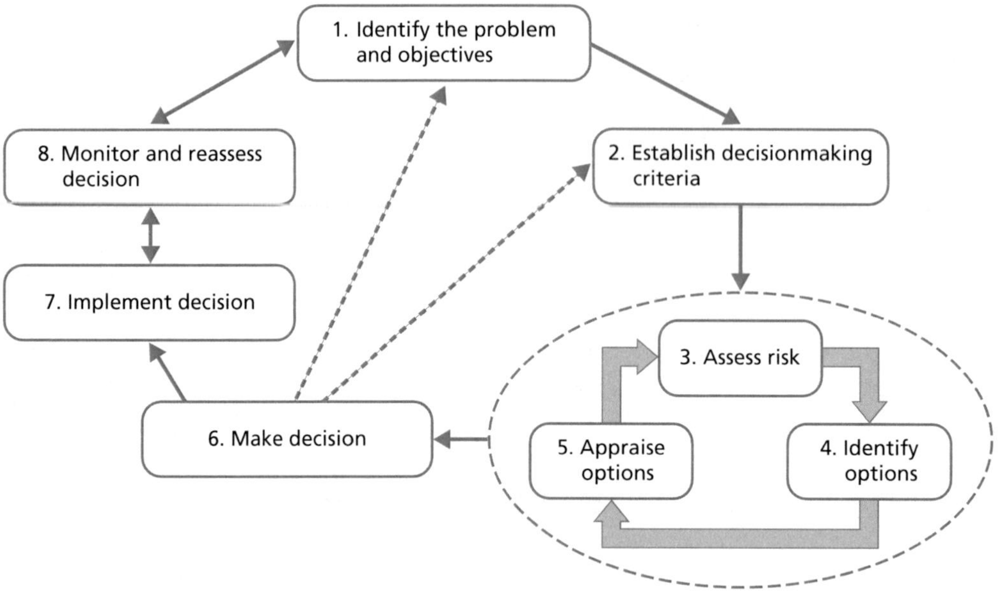

SOURCE: Figure adapted from National Academies of Science (2011).
**RAND** *RR242-2.1*

## Addressing Uncertainty in Long-Term Water-Management Planning Using Robust Decision Making

Embedded in the above approach is the recognition that any robust plan that addresses climate change will need to adapt over time—that deep uncertainty in the future means that no one plan set in place today will be optimal and that the plan cannot be static. There is, however, no single accepted approach for assessing risk, identifying options, appraising options, and then making a decision based on this information (Steps 3–6).

One approach designed to address this need—and the one used as part of the Colorado River Basin Study—is Robust Decision Making (RDM). RDM provides a systematic and objective approach for developing management strategies that are more robust to uncertainty about the future (Groves and Lempert, 2007; Lempert, Popper, and Bankes, 2003). This approach has been used in a number of real-world applications, including energy resources, flood risk management, national defense, and water management (Popper et al., 2009; Dixon et al., 2007; Lempert and Groves, 2010).[1] When applied to water supply planning, RDM helps water managers iteratively identify and evaluate *robust* strategies—those that perform well in terms of management objectives over a wide range of plausible futures but that may perform less well under an assumption that one future may be most likely to occur. Trading off optimality for adequacy across many possible conditions is referred to as "satisficing" (Simon, 1956).

Often, the robust strategies identified using RDM are *adaptive* (as opposed to static), meaning that they are designed to evolve over time in response to new information. RDM helps decision makers identify strategies—including both near-term and deferred decisions or

---

[1]    Information on RDM and applications can be found at the RAND RDMlab website (www.rand.org/rdmlab).

investments—that are shown through the analysis to be effective over a wide range of plausible future conditions. RDM also can be used to facilitate group decision making in contentious situations where parties to the decision have strong disagreements about assumptions and values (Groves and Lempert, 2007; Lempert and Popper, 2005).

The engine that makes RDM run is a sophisticated set of statistical and software tools embedded in a process of participatory stakeholder engagement. RDM helps resource managers develop adaptive strategies by iteratively evaluating the performance of proposed options against a wide array of plausible futures, systematically identifying the key vulnerabilities of those strategies,[2] and using this information to suggest responses to the vulnerabilities identified (Lempert and Collins, 2007; Lempert, Popper, and Bankes, 2003; Means et al., 2010). Successive iterations develop and refine strategies that are increasingly robust. Final decisions among strategies are made by considering a few robust choices and weighing their remaining vulnerabilities.

RDM follows an iterative and interactive series of steps consistent with the "deliberation with analysis" decision support process described by the National Research Council (2009). As shown in Figure 2.2, the process shares many similarities with the National Academies of Science framework and can be used to implement Steps 1-6 (Figure 2.1).

### Structuring Decisions

The first step in RDM is a pure deliberation step—one in which the participants to the decision that needs to be made work together to structure that decision. This step is used to frame the problem by understanding the options possible and the key uncertainties such strategies

**Figure 2.2**
**Steps in Robust Decision Making Process**

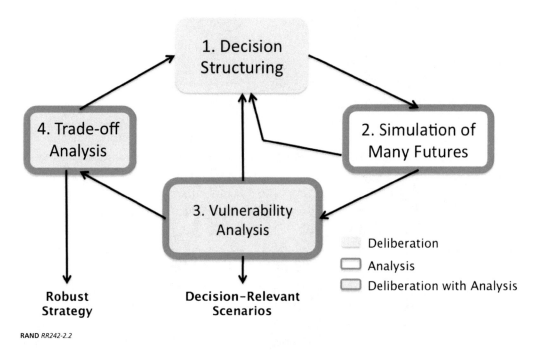

RAND RR242-2.2

---

[2] The approach to identifying key vulnerabilities uses statistical "scenario discovery" algorithms (Bryant and Lempert, 2010; Groves and Lempert, 2007). The terms "scenario discovery" and "vulnerability analysis" are synonymous.

will confront, along with the metrics that will be used to assess how well the strategy or strategies perform. This involves defining the policy questions and structuring the decision analysis to address them in the next step. RDM often uses a framework called "XLRM" to support the decision structuring activity, where "X" stands for the uncertain factors that are used to develop the uncertain futures; "L" stands for management options (or levers) that define strategies to address the various futures; "R" is the relationships among these elements that are reflected in the planning models; and "M" consists of the performance metrics that are used to evaluate and compare management strategies (Lempert, Popper, and Bankes, 2003). In water-planning applications, XLRM provides the information needed to organize the simulation modeling in the next step that captures the response of the water-management system to external conditions related to, for example, future climate, economics, regulatory requirements, and demand projections. The end result of this step is the development of that decision-framing information, which is passed along to the next step.

### Simulation of Many Futures

A key difference between RDM and the typical predict-then-act decision analysis approach is that RDM seeks to evaluate the broadest range of plausible future outcomes without an initial focus on their likelihood. Instead, Step 2 evaluates the baseline and alternative strategies under an expansive set of plausible assumptions about future conditions. This step generates a large database of cases—inputs defining different plausible future conditions and management strategies, coupled with the model-simulated results for outcomes of interest.

Predicting the unpredictable often just leads to bias and gridlock and does not bring managers closer to understanding the merits of their strategy or strategies. Using simulating models to instead define outcomes under a broad-range of assumptions about the future is increasingly considered best practice in climate-change planning and decision support (National Academies of Science, 2011; Lempert et al., 2013).

### Vulnerability Analysis

In Step 3, analysts and decision makers "mine" the database of simulation results (or cases), using visualizations and vulnerability analysis to explore the case results and identify the key combinations of future conditions where one or more candidate strategies might not meet planning objectives. This analysis provides concise descriptions of the combinations of future conditions—what are called "decision-relevant scenarios"—that would make a strategy vulnerable to not meeting its objectives. Such decision-relevant scenarios focus decision makers' attention on the uncertain future conditions most important to the challenges they face and help facilitate discussions about the best ways to respond to those challenges (Bryant and Lempert, 2010; Groves and Lempert, 2007). In the Basin Study analysis, we refer to these as *vulnerable conditions*.

Importantly, this step does not address which of these conditions are more or less *likely* to occur. There remains substantial uncertainty and disagreement regarding how supply and demand conditions on the Colorado will change over time, for example, and the uncertainty is sufficiently deep that it is difficult to estimate the probability of each set of outcomes occurring. Of course, the probability of different outcomes remains an important factor when considering different investment decisions, but consideration of probabilities is deferred until alternative strategies have been defined and compared.

Such vulnerability analysis is a "discovery process" for decision makers, and it is a key feature of RDM. It is most useful in situations in which some combinations of uncertain factors are significantly more important than others in determining whether a strategy meets its goals. In such situations, the analysis can help decision makers recognize those combinations of uncertainties that require their attention and those they can more safely ignore. This information can be useful in itself—shown by the outbound arrow from Step 3 in Figure 2.2—or it can be useful in helping to generate new, more robust strategies to mitigate those vulnerabilities—the iterative arrow that returns to Step 1. Appendix A provides a specific example of how this vulnerability analysis works.

**Trade-Off Analysis**

We often use RDM to do more than just make decision makers aware of the vulnerabilities of a strategy or strategies. Instead, we use the information on potential vulnerabilities as the foundation for evaluating potential modifications of a proposed strategy that might reduce these vulnerabilities (Step 4). RDM supports this step through the use of interactive visualizations that help decision makers and stakeholders see how the system would perform in different futures—particularly those within the vulnerable conditions—under the proposed or augmented strategy. This information is paired with additional information about costs and other impacts of strategies, so that meaningful deliberations over different strategies can occur.

At this point—when deliberating about key trade-offs among different strategies—the decision makers and stakeholders can bring in their assumptions regarding the likelihoods of the vulnerable conditions. For example, if the vulnerable conditions are deemed very unlikely, then the reduction in the corresponding vulnerabilities may not be worth the cost or effort. On the other hand, the vulnerable conditions identified may be viewed as plausible or very likely, lending support for a strategy designed to reduce these vulnerabilities. Finally, if there is substantial disagreement about the likelihood, the strategy can be modified to add adaptivity— that is, to monitor key inputs to the vulnerable conditions and defer or trigger some choices based on observable outcomes over time.

Based on this trade-off analysis, decision makers may choose a robust strategy (the outward arrow in Figure 2.2), or at least some elements of a robust strategy and begin implementation. They may also decide that none of the strategies under consideration is sufficiently robust and return to the decision structuring step (the arrow back to Step 1 in Figure 2.2), this time with deeper insight into the strengths and weaknesses of the strategies initially considered.

## How RDM Was Used in the Basin Study

As described in Chapter Three, prior to the introduction of the RDM methodology, the Basin Study had structured an evaluation of Basin imbalances through the simulation of a large set of scenarios (RDM Steps 1 and 2, Figure 2.2). The RDM approach was then used to implement the vulnerability analysis (Step 3), inform the development and evaluation of dynamic portfolios of options (arrow from Step 3 to Step 1, and second iteration of Step 2), and evaluate trade-offs among four different portfolios (Step 4). The Basin Study did not define a specific robust strategy, although it did identify options that would likely be implemented as part of a robust strategy.

## Summary

This chapter motivated the use of, and described a new methodology for, planning under deep uncertainty—RDM. This iterative, analytic approach can help water planners develop adaptive water-management strategies that are robust to a wide range of plausible but uncertain future conditions. RDM was used in the Basin Study to structure an evaluation of vulnerabilities and trade-offs among portfolios of water-management options.

# CHAPTER THREE

# Developing Robust Management Strategies for the Colorado River Basin

Reclamation and the Basin States will face many challenges managing the Colorado River over the coming decades. Uncertainty about future conditions, the plethora of different and sometimes-competing objectives, and the diversity of options for addressing potential supply and demand imbalances all complicate long-term planning. In January 2010, the Basin Study began to address these challenges by taking a scenario-planning approach to consider how changes in supply and demand might affect future imbalances. With RAND's involvement starting in January 2012, this approach was expanded to include many elements of RDM to help structure the analysis of options and strategies to alleviate the possible Basin imbalances.

This chapter uses the XLRM framework introduced in Chapter Two to describe the decision-structuring step of RDM (Step 1 of Figure 2.2) as applied in the Basin Study. Specifically, it summarizes scenarios developed to reflect future uncertainty (X), the model used to simulate future conditions (R), the subset of performance metrics (M) used to evaluate robustness of the system, and the options (L) and potential robust strategies.

Table 3.1 summarizes the XLRM framework as described in the subsequent sections.

**Table 3.1**
**Summary of Uncertainties, Policy Levers, Relationships, and Metrics Addressed in Study (XLRM Matrix)**

| Uncertainties or Scenario Factors (X) | Management Options and Strategies (L) |
|---|---|
| Demand for Colorado River water<br>Future streamflow or water-supply climate drivers<br>Reservoir operations post-2026 | Current Management<br>Four portfolios composed of individual options<br>• Demand reduction<br>• Supply augmentation |
| **Relationships or Systems Model (R)** | **Performance Metrics (M)** |
| Colorado River Simulation System (CRSS) | Upper Basin Reliability – Lee Ferry Deficit<br>Lower Basin Reliability – Lake Mead Pool Elevation<br>Cost of option implementation |

NOTE: A larger set of performance metrics were evaluated and described in the Basin Study.

## Scenarios and Uncertainties (X)

The Basin Study Team developed a set of supply, demand, and reservoir-operations scenarios during the first year of the study (Reclamation, 2012a).[1] Each scenario describes one plausible way that each of these three factors could evolve over time. Each supply scenario represents a different source of data from which to define future streamflow estimates, and comprises many different 49-year (2012–2060) time series of streamflows that are consistent with the source data's statistical properties. Each of these time series is called a *future trace* or *trace*. Note that there are many other uncertainties about the future system—for example, future instream flow requirements—that were not evaluated in this study.

To support the vulnerability analysis, we characterized each trace by statistics of streamflow and demand. For some traces, we also characterized the climate conditions associated with the projected streamflows.

### Demand Scenarios

The Basin Study Team worked with the Basin Study state partners, environmental nongovernmental organizations, and tribal organizations to develop six demand scenarios (Reclamation, 2012c). These scenarios span a range of plausible future demands, not considering additional programs and incentives for water conservation, and were based on four storylines (Table 3.2). As input to the vulnerability analysis, we calculated the average demand in the last two decades of each climate and demand trace (2041–2060). The post-2040 demand ranges from 13.8 maf (Slow Growth demand scenario) to 15.6 maf (Rapid Growth demand scenario).

Table 3.2
**Demand Scenario Storylines, Scenarios, Descriptions, and Statistics**

| Storyline and Scenario | Description | Average Demand, 2041–2060 (percentage of 2012 baseline) |
|---|---|---|
| Current Projected [CRBS A][a] | Continuation of long-term trends in growth, development patterns, and institutions | 14.4 maf per year (109%) |
| Slow Growth [CRBS B] | Slow growth with emphasis on economic efficiency | 13.8 maf per year (105%) |
| Rapid Growth [CRBS C1 and CRBS C2][b] | Economic resurgence (population and energy) and current preferences toward human and environmental values | 15.6 maf per year (118%) [C1] 14.7 maf per year (111%) [C2] |
| Enhanced Environment [CRBS D1 and CRBS D2] | Expanded environmental awareness and stewardship with growing economy | 14.1 maf per year (107%) [D1] 15 maf per year (114%) [D2] |

[a] The Colorado River Basin Study (CRBS) demand scenario names are provided in brackets.

[b] Basin Study scenarios C1, C2, D1, and D2 represent modest variations to the associated storyline.

NOTE: The reference 2012 baseline (13.2 maf) is set from the current trends estimate of 2012 demand. The demands in this table are valid for the Historical, Paleo, and Paleo/ Historical blend supply scenarios. For the Future Climate supply scenario, demand varies by climate. For example, the Current Projected demand for the Future Climate scenario ranges between 14.3 maf per year and 15.7 maf per year, with a median demand of 14.8 maf per year.

---

[1]   The Basin Study scenarios are distinct from RDM decision-relevant scenarios, which are defined through the vulnerability analysis. To reduce terminology confusion this report uses the phrase "vulnerable conditions" for the "RDM decision-relevant scenarios."

**Supply Scenarios**

The Basin Study also developed four supply scenarios based on different sources of future streamflow estimates (Table 3.3). The first is based on the recent historical record. Each trace within the Historical scenario is a repeat of the historical record (from 1906 to 2007) with a different starting year.[2] The second and third scenarios are based on streamflow estimates derived from paleoclimatological proxies, such as tree ring data. Each trace is consistent with a subset of years from the paleoclimatological record. The fourth scenario is derived from the projections of future climate conditions from 16 global climate models and three global carbon emissions scenarios. Each trace is derived from downscaled results from a single GCM projection. The supply scenarios are described in detail in Technical Report B of the *Colorado River Basin Water Supply and Demand Study* (Reclamation, 2012b).

To support the vulnerability analysis, each trace was characterized by the following statistics of streamflow at Lees Ferry:[3]

- annual mean (2012–2060)
- trend (2012–2060)
- variance (2012–2060)
- annual mean of driest N-year period (examined five-, eight-, and ten-year periods)
- year of driest N-year period (examined five-, eight-, and ten-year periods)
- annual mean flow of wettest N-year period (examined five-, eight-, and ten-year periods)
- year of wettest N-year period (examined five-, eight-, and ten-year periods)

**Table 3.3**
**Supply Scenario Description and Number of Traces**

| Supply Scenario | Number of Traces | Description |
|---|---|---|
| Historical [CRBS Observed Resampled] | 103 | Future hydrologic trends and variability are approximately similar to the past century |
| Paleo [CRBS Paleo Resampled] | 1,244 | Future hydrologic trends and variability are represented by reconstructions of streamflow for a much longer period in the past (nearly 1,250 years) that show expanded variability |
| Paleo/Historical Blend [CRBS Paleo Conditioned] | 500 | Future hydrologic trends and variability are represented by a blend of the wet-dry states of the longer paleo-reconstructed period (nearly 1,250 years), but magnitudes are more similar to the observed period (about 100 years) |
| Future Climate [CRBS Downscaled GCM Projected] | 112 | Future climate will continue to warm with regional precipitation and temperature trends represented through an ensemble of future downscaled GCM projections |

NOTES: *Colorado River Basin Water Supply and Demand Study* supply scenario names are provided in brackets. The assumptions, methods, and results for each of the water-supply scenarios are discussed in detail in Technical Report B (Reclamation, 2012b).

---

[2]  This procedure of developing a set of offset sequences is called the Indexed Sequential Method (ISM).

[3]  Lee Ferry is close to, though slightly downstream from, the U.S. Geological Survey flow gauge at Lees Ferry, Arizona. The Paria River enters the Colorado River between these locations, leading to small differences in flows between the two points. In this report, we use "Lee Ferry" when referring to the Compact delivery requirements from the Upper to Lower Basin, and "Lees Ferry" when referring to natural streamflow measurements of the Colorado River.

Figure 3.1 shows that the nature of each supply scenario varies significantly. The Future Climate scenario, for example, has the widest range of annual river flows, but the lowest median flow (13.5 maf). The Paleo scenario has lower mean flows but also lower variability than the Historical scenario. The Paleo/Historical Blend scenario shows variability similar to the Historical scenario, but with more low-flow traces consistent with the Paleo scenario.

To better understand how climate conditions are related to Colorado River flows and support the vulnerability analysis described in Chapter Four, we also characterized each trace with respect to the temperature and precipitation for the Historical and Future Climate scenarios:[4]

- change in temperature from 1950 to 1999 historical baseline (degrees F)
- change in precipitation from 1950 to 1999 historical baseline (percent)
- variance in temperature and precipitation (2012–2060)
- annual precipitation in wettest period (examined five-, eight-, and ten-year periods)
- annual temperature in hottest period (examined five-, eight-, and ten -year periods).

**Figure 3.1**
**Distribution of Annual Colorado River Streamflow (2012–2060)**
**for Each Basin Study Supply Scenario**

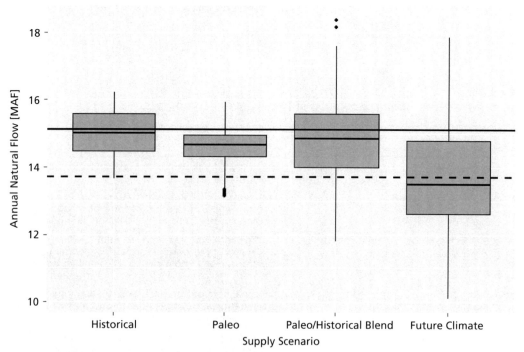

NOTES: Flows correspond to those at Lees Ferry. Solid horizontal line indicates the estimated average flow over the period of record from 1906 to 2007 (15 maf) and the dashed horizontal line indicates the average flow between 1991and 2010 (13.7 maf).
RAND RR242-3.1

---

[4] Unfortunately, traces in the Paleo and Paleo/Historical blend scenarios were developed from paleo reconstructions of streamflow only, and associated paleo-derived climate data were not available for this analysis (Reclamation, 2012b, 2007; Meko et al., 2007). As a result, the climate analysis was performed after the Basin Study was published.

Figure 3.2 shows projected changes in temperature and precipitation for each trace within the Historical and Future Climate scenarios. As expected, the Historical scenario climate traces show only modest positive and negative trends, reflecting which years of the longer historical record are included in each specific 50-year trace. The Future Climate scenario traces, however, show significant warming trends in all traces—most between roughly 2.5 and 4 degrees Fahrenheit. The Future Climate scenario traces show a much larger range in precipitation trends, both increasing and decreasing. Note that the declines in streamflow for the Future Climate scenario traces shown in Figure 3.1 are consistent with traces in which precipitation increases. This is due to increased losses in supply from greater evaporation and plant evapotranspiration under warmer future climate conditions.

**Reservoir Operations Scenarios**

Lastly, two reservoir-operations scenarios reflect different assumptions about how the system would be operated beyond 2026, when the Interim Guidelines are scheduled to expire. In one, the guidelines for Lower Basin shortage allocation and reservoir management are extended; in the other, they instead revert to the "No Action" Alternative as stipulated in the 2007 Interim Guidelines Environmental Impact Statement (EIS) (Reclamation, 2007). For the analysis presented here, continuation of the Interim Guidelines means the continuation of mandatory, agreed-upon Lower Basin shortages to help maintain storage in Lake Mead if the lake eleva-

**Figure 3.2**
**Summary of Temperature Changes Across Traces (left) and Precipitation Changes Across Traces (right) for the Historical and Future Climate Scenarios**

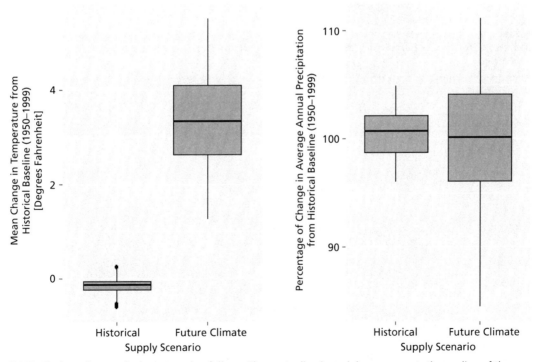

NOTE: The box plots can be interpreted as follows: The center line in each box represents the median of the distribution, which the box ranges from the 25th percentile to the 75th (interquartile range, or IQR) in each set of traces. The lines extending from the boxes reach to 1.5 times the IQR, and any points beyond these cutoffs are displayed as individual outliers.

RAND RR242-3.2

tion drops below 1,075 feet above mean sea level (msl). For those scenarios in which reservoir management switches to the No Action Alternative, the shortage guidelines instead revert to pre-2007 shortage guidelines, which often lead to more severe curtailments for users with junior water rights in the Lower Basin.

### Simulations to Support Baseline Analysis

When evaluating the performance of the Colorado River Basin system, the supply, demand, and post-2026 operations scenarios were combined, as shown in Table 3.4, and totaled 23,508 individual traces. For purposes of the vulnerability analysis, the results were resampled so that each supply scenario would contribute one-quarter of the total number of cases.[5]

## Simulating the Colorado River System (R)

The CRSS is Reclamation's long-term planning model and was used to simulate the Colorado River system for the Basin Study. CRSS simulates operations at a monthly time-step and is developed in the RiverWare® modeling software (Zagona et al., 2001). CRSS models 12 reservoirs (Fontenelle, Flaming Gorge, Starvation, Taylor Park, Blue Mesa, Morrow Point, Crystal, Navajo, Powell, Mead, Mohave, and Havasu), using unique operational rules. Key inputs to a simulation include prescribed monthly streamflows at 29 locations throughout the Basin, schedules for water demand by the Basin's water-using entities (of which there are more than 400), minimum-flow requirements at various parts of the River, and rules for the operation of its reservoirs. CRSS calculates results for hundreds of different *system response variables* to

**Table 3.4**
**Experimental Design for Evaluating the Vulnerabilities of the Baseline Management Approach and with Portfolios**

| Supply Scenarios | Traces | (multiplied by) | Demand Scenarios | (multiplied by) | Post-2026 Operation of Lakes Powell and Mead | (equals) | Traces (percentage of total) |
|---|---|---|---|---|---|---|---|
| Historical (Observed Resampled) | 103 | | 6 | | 2 | | 1,236 (5%) |
| Paleo (Paleo Resampled) | 1,244 | | 6 | | 2 | | 14,928 (64%) |
| Paleo/Historical Blend (Paleo Conditioned) | 500 | | 6 | | 2 | | 6,000 (25%) |
| Future Climate (Downscaled GCM Projected) | 112 | | 6 | | 2 | | 1,344 (6%) |
| Total | 1,959 | | 6 | | 2 | | 23,508 |

---

[5]   For the vulnerability analysis, the Historical, Paleo/Historical Blend, and Future Climate scenarios were replicated so that they were the same size as the Paleo scenario (14,928 records). For the analysis of options, the target sample size was 6,000, and the Paleo results were randomly sampled down to this level whereas the other scenarios were replicated to match this sample size.

depict system operation and response to varying hydrologic, demand, and operating criteria. Figure 3.3 shows a simplified schematic of the CRSS network. The black line represents the model schematic of the Colorado River and its tributaries. The red and purple symbols indicate major demands represented by the model. The colored regions denote individual basins that

**Figure 3.3**
**Simplified Schematic of CRSS Network**

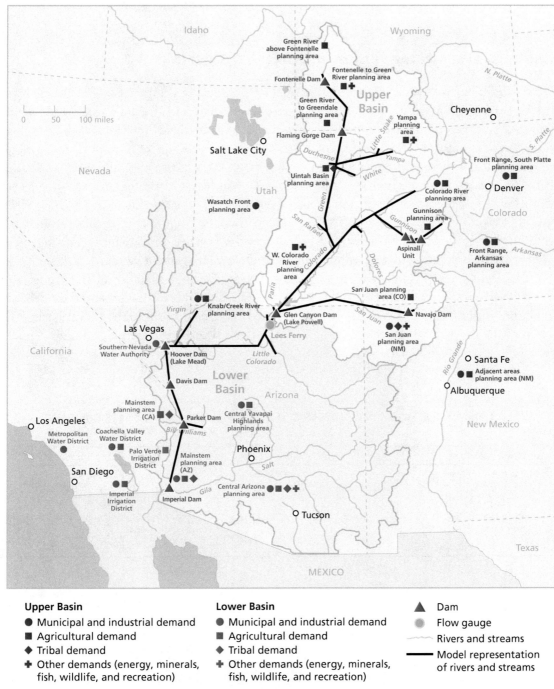

**Upper Basin**
● Municipal and industrial demand
■ Agricultural demand
◆ Tribal demand
✚ Other demands (energy, minerals, fish, wildlife, and recreation)

**Lower Basin**
● Municipal and industrial demand
■ Agricultural demand
◆ Tribal demand
✚ Other demands (energy, minerals, fish, wildlife, and recreation)

▲ Dam
● Flow gauge
〜 Rivers and streams
━ Model representation of rivers and streams

contribute supply to the Basin. The light blue line represents the actual route of the Colorado River and its tributaries across the seven-state region.

CRSS was used to evaluate the complete set of these traces in Table 3.4 many times for this study. First, the traces were evaluated to identify the vulnerabilities of the baseline management of the system (described in Chapter Four). Next, they were evaluated for each of the four portfolios discussed above (described in Chapter Five). The computing resources to perform these simulations and process the model outputs were extensive: Computing facilities were used at both the Center for Advanced Decision Support for Water and Environmental Systems at the University of Colorado, Boulder, and the RAND Corporation.

## Metrics to Evaluate the Performance of the System (M)

The Basin Study evaluated the performance of the system using a large set of system-reliability metrics corresponding to six resource categories—water deliveries (nine metrics), electric power resources (two metrics in three locations), water quality (one metric in 20 locations), flood control (three metrics in ten locations), recreational resources (two metrics in 13 locations), and ecological resources (five metrics in 34 locations) (Reclamation, 2012g). A smaller subset of metrics was then developed to provide a high-level view of system resources under different traces.

This report focuses on two water delivery metrics—Lee Ferry Deficit and Lake Mead Pool Elevation—shown in Table 3.5.[6] We refer to these as *performance metrics* through the remainder of this discussion. This report also uses the costs of implementing management options as an additional measure of performance for the management system.

The two system-reliability metrics broadly summarize the reliability of the Upper and Lower Basins, respectively. If there is a Lee Ferry deficit, then there is the potential for delivery reductions in the Upper Basin to augment flows from the Upper Basin to the Lower Basin. The health of the Lower Basin system and deliveries to the Lower Basin states are similarly closely tied to the Lake Mead elevation.

Table 3.5
**Performance Metrics and Vulnerability Thresholds for the Two Water-Deliveries Metrics Used in This Report**

| Performance Metric | System Reliability Metric | Vulnerability Threshold |
|---|---|---|
| Upper Basin Reliability | Lee Ferry Deficit | Any deficit: running ten-year sum of deliveries falls below 75 maf in any one month |
| Lower Basin Reliability | Lake Mead Pool Elevation | Reservoir level below 1,000 feet msl in any one month |
| Cost of Strategy Implementation | n/a | n/a |

---

[6]  Lee Ferry, Arizona, is considered the dividing line between the Upper and Lower Basins in the Compact (Reclamation, 2012g). The Compact states that "...The States of the Upper Division will not cause the flow of the River at Lee Ferry to be depleted below an aggregate of 75,000,000 acre-feet for any period of ten consecutive years" (67th Congress, 1923). Such a depletion has never occurred, but it is generally agreed that deliveries to the Upper Basin would need to be curtailed—via a "Compact Call"—if this provision were threatened by low flows over a ten-year period. A "Lee Ferry Deficit" is the amount of curtailment that would be necessary in such an event. The Basin Study used the methods described in this report to look at six primary water delivery metrics.

For the two system-reliability metrics, the Basin Study further identified thresholds of performance beyond which Upper or Lower Basin water deliveries become vulnerable. These thresholds define management goals for the two Basins. For example, a relevant threshold for Lake Mead Pool Elevations is 1,000 feet—the level below which the Southern Nevada Water Authority can no longer withdraw water using its lowest current intake. Figure 3.4 illustrates two example simulations of Lake Mead Pool Elevation over time. If the reservoir elevation falls below the 1,000-foot threshold, the Colorado River management system does not meet its baseline water delivery objectives.

For the Upper Basin, the existence of a Lee Ferry Deficit metric, defined as any time that the running ten-year sum of deliveries past Lee Ferry falls below 75 maf in any single month, signals that *Upper Basin Reliability* goals are not met. For Lake Mead levels, any simulation in which the reservoir level drops below 1,000 feet signals that *Lower Basin Reliability* goals are not met.

## Options and Strategies to Improve Performance (L)

The Basin Study evaluated the baseline reliability of the Colorado River system by simulating current operating rules and procedures. This included the water allocations to users in the Basin States per the Law of the River. It represents the operational rules currently in place to balance reservoir storage between Lakes Powell and Mead. It also reflects the Interim Guidelines agreement on how to curtail deliveries during shortage periods. As the Interim Guidelines will expire in 2026 if no further management changes are made, the scenarios test outcomes under the assumption that the guidelines are extended and the assumption that they are not.

Within this context, the Basin Study evaluated a wide array of different supply-augmentation and demand-reduction options that could improve system performance and reduce vulnerabilities. The Basin Study first solicited the public for input and received propos-

**Figure 3.4**
**Example Simulations in Which Lake Mead Pool Elevation Objectives Are Met and Not Met**

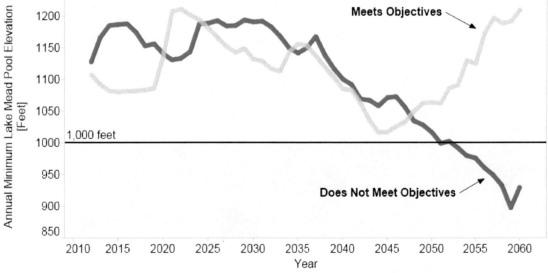

als for about 150 different options. The proposals ranged from broad-based strategies, such as improving water use efficiency in Southern California, to specific schemes for importing water into the Basin through new conveyance facilities. All the options are described in Technical Report F (Reclamation, 2012f).

The Basin Study Team then evaluated a smaller set of these options—about 80—according to rough estimates of option yield, availability, and 16 other criteria, including technical feasibility, permitting risk, legal risk, policy risk, and energy intensity.[7] The Basin Study Team assigned each project a score between A and E for each of the 16 additional criteria, with an A score indicating the most favorable characteristics and E indicating the least favorable. Estimates of option cost were also developed. Appendix B and Chapter Six provide descriptions of the main option types.

## Portfolios of Prioritized Options

The RAND team developed a tool to help define strategies, each defined by a portfolio of specific options. The "Portfolio Development Tool" helped the Basin Study Team develop prioritized lists of options. The Portfolio Development Tool was designed with a simple, user-friendly framework that allowed the Basin Study Team, state partners, and other stakeholders to assimilate the large amounts of information developed by the study on more than 80 different options and develop portfolios for evaluation that were consistent with an underlying strategy.

The Portfolio Development Tool constructed portfolios based on three types of information. First, it evaluated the cost-effectiveness of each option, calculated by dividing the expected supply yield or demand reduction by the total cost of the project.[8] Available options were prioritized by cost-effectiveness—the most cost-effective options were to be implemented first. Next, the Portfolio Development Tool excluded options that did not meet user-specified limits for the qualitative scores (A–E) developed for each criterion. Lastly, the Portfolio Development Tool adjusted the prioritized list of options each year, based on option availability—options not yet available for implementation were excluded from each year's list. The tool, for example, could be used to develop a portfolio that excludes options with low qualitative scores for energy use or permitting risk and would prioritize all other options by cost-effectiveness.

Figure 3.5 shows the complete prioritized list of options for one portfolio developed for the Basin Study. The options are ordered by cost-effectiveness (text labels). The yield of each option is proportional to the size of the symbol; the position of each symbol along the horizontal axis indicates when each option could become available. As described below, CRSS implements the top available option not yet implemented when one or more signposts indicate the need for additional management. The actual order of implementation may not follow this list, therefore, as different options are available at different times. See Appendix B for details about the individual options.

The Portfolio Development Tool was used by the Basin Study Team and stakeholders to develop four portfolios, named A-D in the *Colorado River Basin Water Supply and Demand*

---

[7]   The study team was unable to analyze each option individually using more detailed criteria for the purposes of prioritizing options within a portfolio. As described below, however, the portfolios are evaluated not on these criteria, but on their ability to collectively reduce vulnerabilities.

[8]   The study team developed estimates for the expected yield and total cost for each project during the compilation and characterization of the options described above.

**Figure 3.5**
**Prioritized List of Options for One Portfolio (*Portfolio A*)**

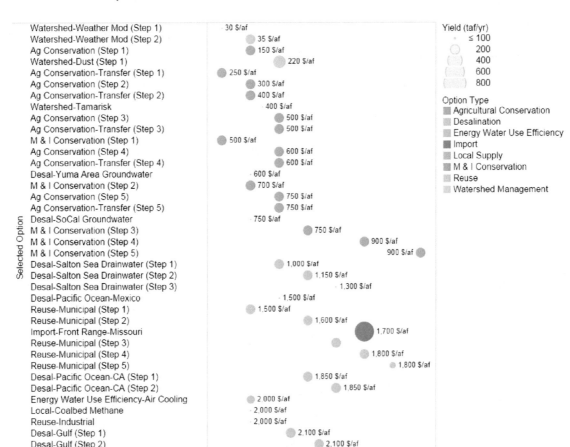

*Study* reports (Reclamation, 2012f, 2012h). To improve the readability of this report, we add descriptive phrases to each portfolio name—*Portfolio A (Inclusive), Portfolio B (Reliability Focus), Portfolio C (Environmental Performance Focus),* and *Portfolio D (Common Options).*

The first two portfolios that were developed focused on different types of options for addressing vulnerabilities. *Portfolio B (Reliability Focus)* included options that were viewed as well-understood and time-tested, and that would provide reliable supplies if implemented. It includes options that cost less than $2,500 per af per year and excludes options that were assumed to have high permitting, legal, and policy risks.

*Portfolio C (Environmental Performance Focus)* is less restrictive in terms of long-term viability, but is more restrictive in terms of most environmental and social criteria. It excludes options that required significant energy use, such as long-distance interbasin transfers and large seawater desalination projects. It includes an Upper Basin water bank project designed to store excess water in the Upper Basin in wet years to support more consistent river flow during dry periods. It also includes options that cost less than $4,200 per af per year and excludes options that were assumed to have high permitting, legal, and policy risks.

Two additional portfolios were then crafted based on these portfolios. *Portfolio A (Inclusive)* included all options included in *Portfolio B (Reliability Focus)* and *Portfolio C (Environmental Performance Focus)*. Alternately, *Portfolio D (Common Options)* included only those options common to both. Figure 3.6 illustrates option commonalities among portfolios and Table 3.6 provides descriptions of each portfolio.

If all options in *Portfolio A (Inclusive)* were implemented, 6.3 maf per year of new supply or reduced demand would be realized (Table 3.7). The main differences between *Portfolio B (Reliability Focus)* and *Portfolio C (Environmental Performance Focus)* are that the former includes more desalination (1.5 maf per year versus 0.6 maf per year) and imported supply options (0.6 maf per year versus 0 maf per year) and the latter includes more watershed management options and an Upper Basin water bank (not represented in the table, as it does not increase yield of the system—it only reallocates it).

**Figure 3.6**
**Schematic Representing Options Included in Each of the Four Portfolios**

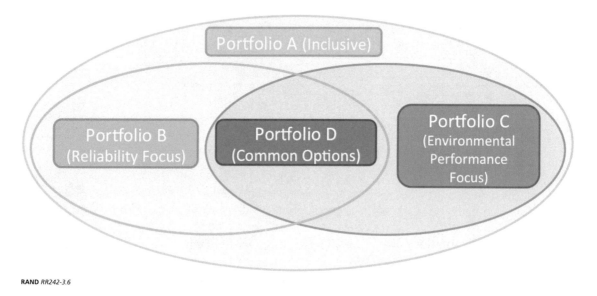

RAND *RR242-3.6*

**Table 3.6**
**Descriptions of Four Portfolios**

| Portfolio Name | Portfolio Description |
|---|---|
| *Portfolio A (Inclusive)* | Includes all options included in the other portfolios |
| *Portfolio B (Reliability Focus)* | Emphasizes options with high technical feasibility and high long-term reliability; excludes options with high permitting, legal, or policy risks |
| *Portfolio C (Environmental Performance Focus)* | Excludes options with relatively high energy intensity; includes options that result in increased instream flows; excludes options that have low feasibility or high permitting risk |
| *Portfolio D (Common Options)* | Includes only those options common to the *Portfolio B (Reliability Focus)* and *Portfolio C (Environmental Performance Focus)* |

NOTE: The portfolio names presented in this table were developed for this report only. The names in parentheses were the names used in the *Colorado River Basin Water Supply and Demand Study.*

**Table 3.7**
**Summary of Options Included in the Four Portfolios**

| Option Category | Option Type | Portfolio A (Inclusive) | | Portfolio B (Reliability Focus) | | Portfolio C (Environmental Performance Focus) | | Portfolio D (Common Options) | |
|---|---|---|---|---|---|---|---|---|---|
| | | Number of Options | Sum of Yield (taf/year) | Number of Options | Sum of Yield (taf/year) | Number of Options | Sum of Yield (taf/year) | Number of Options | Sum of Yield (taf/year) |
| Increase Supply | Desalination | 10 | 1,476 | 10 | 1,476 | 5 | 620 | 5 | 620 |
| | Import | 1 | 600 | 1 | 600 | 0 | 0 | 0 | 0 |
| | Local Supply | 2 | 175 | 1 | 100 | 1 | 75 | 0 | 0 |
| | Reuse | 7 | 730 | 6 | 972 | 7 | 1,150 | 6 | 972 |
| | Watershed Management | 5 | 730 | 2 | 300 | 5 | 730 | 2 | 300 |
| | **Total** | **25** | **4,131** | **20** | **3,448** | **18** | **2,575** | **13** | **1,892** |
| Reduce Demand | Agricultural Conservation | 5 | 1,000 | 5 | 1,000 | 5 | 1,000 | 5 | 1,000 |
| | Energy Water Use Efficiency | 1 | 160 | 1 | 160 | 1 | 160 | 1 | 160 |
| | Municipal and Industrial (M&I) Conservation | 5 | 1,000 | 5 | 1,000 | 5 | 1,000 | 5 | 1,000 |
| | **Total** | **11** | **2,160** | **11** | **2,160** | **11** | **2,160** | **11** | **2,160** |
| | **Grand Total** | **36** | **6,291** | **31** | **5,608** | **29** | **4,735** | **24** | **4,052** |

NOTES: *Portfolio A (Inclusive)* and *Portfolio C (Environmental Performance Focus)* exclude any options that cost more than $4,200 per af. *Portfolio B (Reliability Focus)* and *Portfolio D (Common Options)* exclude any options that cost more than $2,500 per af.

### Dynamic Portfolios to Alleviate Vulnerabilities

To evaluate how each portfolio of options would perform across the wide range of futures, the Basin Study Team defined rules within CRSS to implement options only when conditions indicated a need for them.

To implement these *dynamic portfolios*, the RAND team developed a set of signposts that would trigger the implementation of an option during a CRSS simulation. Signposts were developed for each of the six water delivery indicator metrics (Table 3.8). Signposts specify a set of observable system conditions and thresholds that indicate vulnerabilities are developing. During a simulation, CRSS monitors the signpost conditions: If any thresholds are crossed, options are implemented from the top of the portfolio option list.[9]

This dynamic implementation of portfolios seeks to more realistically mimic how options would be implemented over time in response to system needs. Figure 3.7 illustrates how a simple dynamic portfolio that includes a signpost based on Lake Mead elevation would alleviate an impending Lake Mead vulnerability. In this simple example, the signpost triggers the implementation of new options when Lake Mead falls below about 1,100 feet. The yellow line shows that the implementation of new options leads to higher Lake Mead levels than without the new options. As a result, the vulnerability threshold of 1,000 is never crossed.

**Table 3.8**
**Definitions of the Vulnerability Signposts Developed for Each Indicator Metric**

| Vulnerability | Conditions |
|---|---|
| Lee Ferry Deficit | Low Lake Powell levels (<3,490 feet) and low five-year mean flow at Lees Ferry (<12.39 maf) |
| Lake Mead Pool Elevation | Low Lake Mead elevations (<1,075 feet) and low five-year mean flow at Lees Ferry (<13.35 maf) |
| Upper Basin shortages | Upper Basin shortage greater than 25 percent |
| Lower Basin shortages (two types) | Low Lake Mead elevation (between 1,075 and 1,060 feet) and low five-year mean flow at Lees Ferry (<13.51 maf) |
| Demand above lower division states' basic apportionments | Demand above basic apportionments is within 100,000 af of permissible level |

NOTES: Although this report focuses on the Lee Ferry Deficit and Lake Mead Pool Elevation vulnerabilities, the dynamic portfolios evaluated for this study included signposts for all five of the Water Delivery indicator metrics.

## Summary

This chapter described the key elements of the vulnerability and management strategy analysis performed for the Basin Study. It described the demand, supply, and reservoir operations scenarios, the simulation model used to evaluate the performance of the Colorado River Basin system, and the key metrics used to evaluate the robustness of the system. It concluded by describing the set of individual options evaluated by the Basin Study and the Portfolio Development Tool used to develop four different portfolios of options for reducing Basin imbalances. The next chapter presents the results of the analysis of vulnerabilities of the current management approach.

---

[9]  CRSS implements enough options to add 200 thousand acre-feet (taf) per year of new supply or reduce demand by 200 taf/year each year in which a signpost is triggered. CRSS also only implements those options prespecified to be effective at improving the corresponding indicator metric.

**Figure 3.7**
**Simulated Lake Mead Pool Elevation Over Time Without Options (red) and With Options Implemented by a Simple Dynamic Portfolio**

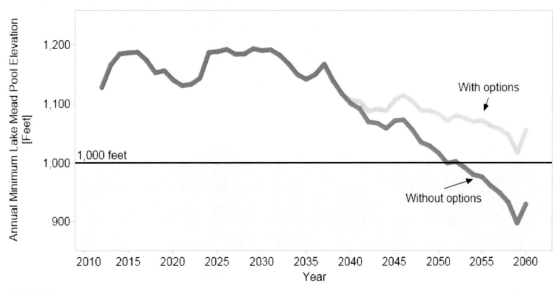

# CHAPTER FOUR

# Future Vulnerabilities to Colorado Basin Water Deliveries

## Introduction

There is substantial uncertainty about future conditions in the Colorado River Basin. As noted in Chapter One, demand for water from the River is expected to increase in both the Upper and Lower Basins, but projected rates of demand growth vary widely. Future streamflow conditions could resemble what we have observed over the last century or what we can infer about the more distant past, but simulations taking into account a changing climate suggest that permanent shifts in future supply are possible.

Given the wide range of outcomes possible across the scenarios considered in this study and discussed in Chapter Three, our goal in this chapter is to first understand which supply and demand conditions most often lead to the Basin not meeting its long-term water delivery objectives. This investigation does not yet include new options to increase supply or reduce demand. Instead, the goal is to concisely describe the external drivers of supply and demand conditions, or *vulnerable conditions*, that most often lead to not meeting objectives under the current management system and with no new investments made.

As such, the analysis presented in this chapter provides a snapshot of what conditions could lead to Basin vulnerability over the 49-year time horizon (2012–2060), and helps decision makers understand the range of conditions for which the current management system continues to meet objectives. As is true of vulnerability analysis in RDM (discussed in Chapter Two), the results of the analysis presented here help focus the decision maker's attention on those conditions that directly affect Basin management decisions—which scenarios matter to the investment decisions Reclamation and the Basin States will be making in future years—and helps facilitate discussions about the best ways to respond to these potential challenges. Importantly, this step does not address which of these conditions are more or less *likely* to occur. The probability of different outcomes remains an important factor when considering different investment decisions, but a discussion of these probabilities is deferred until the actual management strategies are defined and compared in the next chapter.

We begin this chapter by summarizing the future traces under which the Basin does not meet its water delivery objectives across the thousands of traces simulated for the Basin Study (the end product of the simulation of futures step in RDM described in Chapter Two). We use the vulnerability analysis methods described in Step 3 of RDM in Chapter Two to identify a concise set of vulnerable conditions for each of the two long-term delivery reliability metrics in this analysis: Lee Ferry Deficit (Upper Basin Reliability) and Lake Mead Pool Elevation below 1,000 feet msl (Lower Basin Reliability). A Lee Ferry Deficit occurs in the CRSS simulations when the ten-year sum of flows at Lee Ferry falls below 75 maf. An important reliability goal for the Upper Basin states is to avoid crossing this threshold even once in future years. The Lake Mead elevation reaches an unacceptable level when it drops below 1,000 feet elevation (msl), currently the elevation of Southern Nevada Water Authority's lowest intake. Following the approach in the Basin Study's summary *Study Report*, we use this metric as a proxy for the long-term reliability of all Lower Basin water deliveries (Reclamation, 2012h). Crossing this threshold even once over the 2012–2060 period—which would lead to cutting off most of Las Vegas's water supply—would not meet Basin management objectives.

For each metric, conditions are defined using two different descriptions in supply: streamflow volume, which includes all four of the supply scenarios in the analysis, or climate drivers, for only the Historical and Future Climate supply scenarios. These vulnerable conditions are then used as the backdrop against which the portfolios of options discussed in Chapter Three are compared in the next chapter.

## Under Which Futures Does the Basin Not Meet Water Delivery Objectives?

The Basin Study Team evaluated the future performance of the Colorado River Basin system using the CRSS model for the 23,508 simulated traces (see Table 3.4 in Chapter Three). To describe future Basin vulnerability in this sample, we first summarize system performance for the two key water delivery reliability metrics: Lee Ferry Deficit (Upper Basin Reliability) and Lake Mead Pool Elevation below 1,000 feet msl (Lower Basin Reliability; see Table 3.8 in Chapter Three).

Figure 4.1 summarizes outcomes in two ways: (1) the percentage of traces in which management objectives are not met at least once during the time period (left column), and (2) the

**Figure 4.1**
**Summary of Long-Term Water Delivery Outcomes That Do Not Meet Objectives**

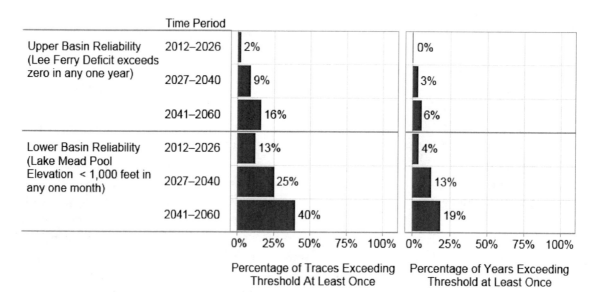

NOTE: For the entire period, Lee Ferry Deficits occur in 19 percent of traces and 4 percent of years, and Lake Mead Pool Elevation falls below 1,000 feet in 47 percent of traces and 13 percent of years.
RAND RR242-4.1

percentage of all years in the simulation in which outcomes did not meet objectives (right column). Rather than provide an average frequency for each trace, the percentage of years summarizes the overall frequency with which the Basin does not meet delivery objectives. This helps to distinguish between, for instance, traces in which the management objectives are not met only once versus traces in which this occurs many times. Figure 4.1 shows these results in three separate time periods over the 49-year horizon: 2012 through 2026, 2027 through 2040, and 2041 through 2060. The end of the first time period was selected to coincide with the expiration date of the 2007 Interim Guidelines (Reclamation, 2007).

Figure 4.1 shows that for Upper Basin Reliability, the percentage of traces in which at least one Lee Ferry Deficit occurs increases from 2 percent in the period 2012 through 2026 to 16 percent in the period 2041 through 2060, with Lee Ferry Deficits occurring in 6 percent of years in the last period (top set of rows). Similarly, for Lower Basin Reliability, Lake Mead Pool Elevations fall below the 1,000-foot threshold more frequently across traces and years in later periods. By the last period, Lake Mead levels are vulnerable in 40 percent of traces and 19 percent of years. Although we see vulnerabilities emerge more frequently in the simulation results, this does not necessarily show that the likelihood of vulnerability is increasing. The apparent increase in vulnerable conditions over time is a consequence of the implicit assumption in our approach that each supply scenario is equally likely, which we make in the absence of more definitive information about the probability distribution. Other assumptions could be made with different implications. For example, if Historical or Paleo supply scenarios were assumed to be more likely, then we would observe little to no increase in vulnerability. This is because water supply trends in those scenarios do not decline as much as in many traces drawn from the Future Climate scenario.

## What Future External Conditions Lead to Vulnerabilities?

Next, we look at each of the two reliability metrics in turn to identify external conditions most often associated with not meeting objectives—what we refer to in the RDM approach as vulnerable conditions. Specifically, we use the vulnerability analysis methods described in Chapter Two and characterizations of the uncertainties described in Chapter Three to identify one or more sets of vulnerable conditions for each metric. The goal is to understand which combination of future uncertain factors and restrictions on their ranges best strike a balance between describing the most outcomes that do not meet the Basin's objectives (what is called coverage), only describing outcomes that do not meet the Basin's objectives (what is called density), and describing the vulnerable conditions in a simple and understandable way. In principle, we would want density and coverage to be 100 percent and to need the least number of combinations of uncertain factors and restrictions on their ranges to explain vulnerable conditions. But in practice, these three factors trade off among one another. Thus, we are looking for the best balance among them.

### Upper Basin Vulnerable Conditions

As noted above, in the study simulations, the Upper Basin experiences a Lee Ferry Deficit in 19 percent of traces by the last 20 years of the simulation. In mining those simulations in the vulnerability analysis, we first sought to answer the question: what future conditions typically lead to Lee Ferry Deficits?

First, using streamflow to characterize future water supply, the Upper Basin is susceptible to a Lee Ferry Deficit in conditions in which long-term average streamflow declines beyond what has been observed in the recent historical record. Specifically, we identified a set of vulnerable conditions—what we referred to above as a combination of future uncertain factors and restrictions on their ranges—called *Declining Supply* that corresponds to long-term average flows at the Lees Ferry, Utah, flow gauge below 13.8 maf per year, coupled with an eight-year period of consecutive drought years where the average flow dips below 11.2 maf per year. Traces that meet both of these conditions—that is, they have low long-term mean flows and an eight-year drought of this magnitude—lead to a Lee Ferry Deficit 87 percent of the time (high density). In addition, *Declining Supply* captures 85 percent of all traces with at least one Lee Ferry Deficit (high coverage). Moreover, it only requires the two input parameters to produce such vulnerable conditions, so it is simple to understand and interpret.

The results of the analysis are summarized visually in Figure 4.2. Each point in the figure represents one trace in the analysis, characterized according to long-term mean annual flow (vertical axis) and mean annual flow during the driest eight-year period (horizontal axis).[1] Red Xs indicate traces with at least one Lee Ferry Deficit during the simulation, and gray Os mark

**Figure 4.2**
*Declining Supply* **Vulnerable Conditions for Lee Ferry Deficit (Streamflow Variables)**

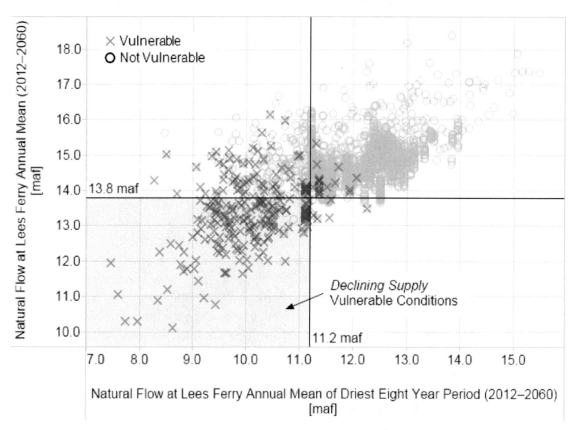

NOTE: This scatterplot shows results from thousands of individual traces, many of which overlap.
**RAND** *RR242-4.2*

---

[1]   This plot shows results from thousands of individual traces, many of which overlap in the scatterplot.

traces in which no such deficit occurs and the Upper Basin meets its objectives. The yellow region (lower left) in the figure summarizes the vulnerable condition boundaries identified in the analysis.

In this example, the Upper Basin experiences a Lee Ferry Deficit in 19 percent of traces, which means that 81 percent of the traces do not lead to a deficit. Thus, the red Xs represent the 19 percent of the traces that lead to a deficit, and the gray Os represent the 81 percent of traces that do not lead to a deficit. The red Xs within the yellow region are those results that lead to a deficit and are described by the *Declining Supply* vulnerable conditions. The *Declining Supply* vulnerable conditions encompasses most but not all of the vulnerable traces, which is why some of the vulnerable traces (red Xs) appear outside of the boundaries.

Figure 4.2 shows that, despite the evident correlation between long-term low flows and acute drought flows, a combination of restrictions in both dimensions is important to identify a set of vulnerable conditions with high coverage and density. For instance, removing the long-term mean restriction would include all traces in the upper-left quadrant. Although there are some traces in this quadrant in which a Lee Ferry Deficit occurs that are not captured in Declining Supply vulnerable conditions, a majority of the traces in this quadrant are not vulnerable, and therefore density would drop dramatically if this region were included.

The *Declining Supply* set of vulnerable conditions has several notable features. First, this set is based only on supply uncertainty and does not depend upon future demand. This does not mean that increasing demand is not an important driver of future vulnerability. Demand rises across all six demand scenarios considered (as discussed in Chapter Three), and the analysis suggests that the exact rate of increase is less important than the fact that the rate is always positive, within the range of future demands considered. Alternately, if the Basin Study were to include a scenario with flat or declining demand, then future demand could emerge as a key part of this set of vulnerable conditions.

Another feature is that these conditions are not present in all Basin Study supply scenarios. A long-term mean of 13.8 maf per year and an eight-year drought minimum of 11.2 maf per year do not occur in the observed resampled (Historical) scenario and have not previously been observed over the roughly 100-year period of record. However, the conditions do emerge in a handful of paleo traces, as well as many more from the blended Historical and Paleo scenario. By definition, this set of conditions is beyond our recent observations, but it can be found in selected low flow and drought periods present in the paleohistorical record, or in nonstationary future climate scenarios in which overall streamflow is declining.

A similar pattern emerges when we use climate inputs instead of streamflow to define a set of vulnerable conditions for the Lee Ferry Deficit metric (Figure 4.3). Here, we identified a set of conditions defined with two climate parameters: positive temperature change in the Upper Basin greater than approximately 2 degrees Fahrenheit over the period 2012–2060 relative to the historical 1950–1999 baseline, and an average Upper Basin future precipitation (2012–2060) that is at or below 100 percent of the 1950–1999 average. In the Historical and Future Climate supply scenarios, 85 percent of traces with these conditions have at least one Lee Ferry Deficit (density), and they capture 92 percent of all traces where objectives are not met (coverage). Once again, only two input parameters are needed to describe such vulnerable conditions, leading them to be easily understood and interpreted.

Figure 4.3 shows a scatterplot of the climate vulnerable conditions identified for the Upper Basin Reliability metric, with the same symbology as Figure 4.2. This plot shows that the climate conditions in the historical set differ substantially from those found in the Future

**Figure 4.3**
**Vulnerable Conditions for Lee Ferry Deficit (Climate Variables)**

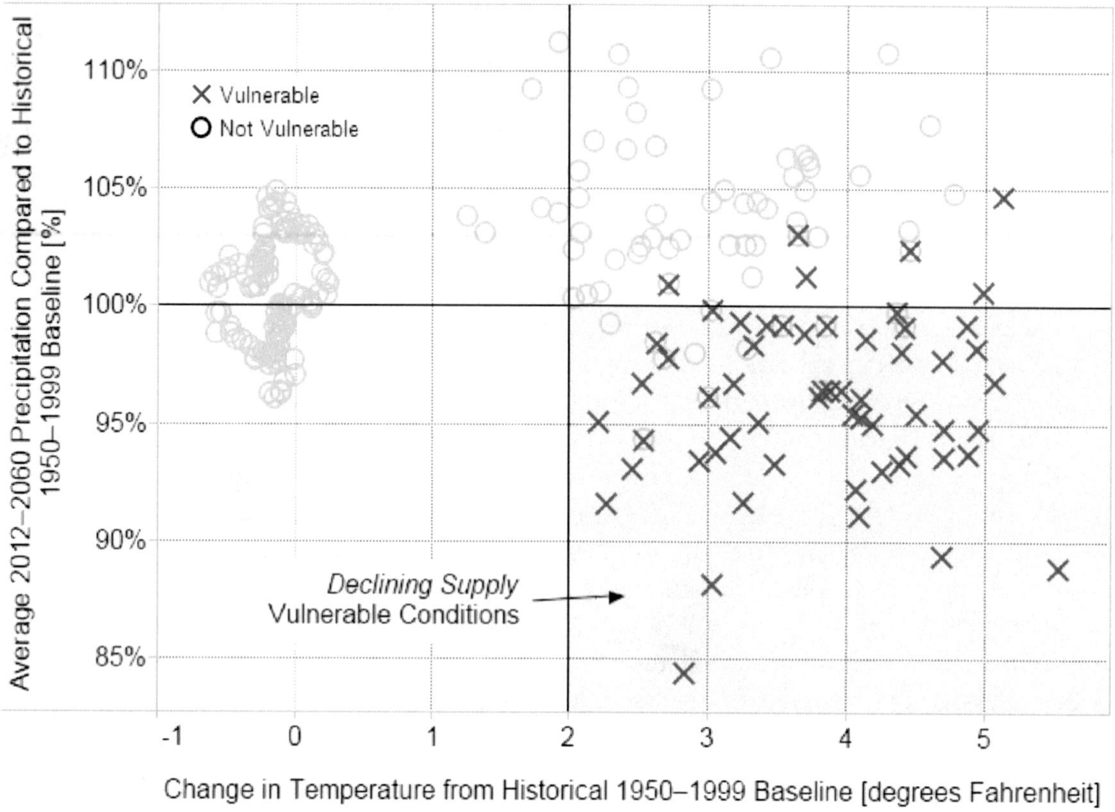

NOTE: This scatterplot shows results from thousands of individual traces, many of which overlap.
RAND RR242-4.3

Climate scenario projections. The tight cluster of points in the left pane of the figure shows the historical set, while the Future Climate projections spread out to the right. The vulnerable conditions only include traces with a positive temperature increase of more than 2 degrees Fahrenheit, which primarily separate the Historical scenario (not included in vulnerable conditions) from the Future Climate scenario (partially included). This is an understandable result, because there are no traces in the Historical Supply scenario where Upper Basin objectives are not met (i.e., no red-colored Xs in the left portion of the figure).

Figure 4.3 also shows that the Upper Basin meets its objectives in some of the Future Climate sequences. The Future Climate scenario traces considered in the Basin Study have long-term average precipitation values at 84 to 111 percent of the historical average. Nearly all Future Climate traces with average precipitation values greater than the historical average (101 to 111 percent)—more than half of the climate traces in total—do not show a deficit at Lee Ferry and meet objectives, regardless of the temperature change projected in that trace (shown on the horizontal axis). This is an important insight and suggests that our ability to place probabilities on future vulnerability will improve dramatically as we grow more confident about precipitation estimates for the Upper Basin from having improved regional climate models.

A major challenge in interpreting the climate vulnerable conditions developed for Upper Basin Reliability is that they do not incorporate paleo scenarios; as a result, these conditions show a major gap in temperature estimates between the historical observed and climate projected traces. The temperature increase restriction used to define vulnerable conditions, estimated at 2 degrees Fahrenheit, is therefore inexact. Thus, additional traces with temperature changes in this range might better determine to which temperatures the Upper Basin is vulnerable.

When considering only the Historical and Climate Supply scenarios, both the streamflow and climate vulnerable conditions identify nearly the same set of traces. Overall, 95 percent of traces are categorized in the same manner—in or out of vulnerable conditions—across the two types of vulnerable conditions. Thus, even though the streamflow conditions were built using different supply scenarios and input parameters, both sets of conditions include nearly the exact same set of vulnerable traces. This is not surprising, because there is a clear causal relationship between temperature, precipitation, and streamflow. As a result, we refer to both definitions as *Declining Supply* and will use them interchangeably through the remainder of the discussion. For convenience and to make best use of the data, the streamflow-derived vulnerable conditions will be referenced in all subsequent figures and numeric results.

The *Declining Supply* vulnerable conditions are summarized in Table 4.1.

## Lower Basin Vulnerable Conditions

Using the same approach we described for the Upper Basin, we defined two sets of vulnerable conditions for the Lake Mead Pool Elevation metric, one using streamflow and one using climate inputs, to describe water supply conditions.

First, using streamflow to characterize water supply, Lower Basin Reliability is vulnerable to conditions in which supplies are simply below the long-term historical average. These Low Historical Supply conditions correspond to the lower half of flows present in the Historical supply scenario as well as many traces from the Paleo, Paleo/Historical blend and Future Climate supply scenarios. Specifically, if the long-term average streamflow at Lees Ferry falls below 15 maf, and an eight-year drought with average flows below 13 maf occurs, a trace is

**Table 4.1**
**Vulnerable Conditions Defined for Lee Ferry Deficit: Declining Supply**

| Streamflow Conditions | | Climate Conditions | |
|---|---|---|---|
| Vulnerable Traces: 19%<br><br>Vulnerability Statistics:<br>• Coverage: 85%<br>• Density: 87% | Vulnerability Definition:<br>• Annual Mean Natural Flow at Lee Ferry (2012–2060) Less Than 13.8 maf:<br><br>10.0　　13.8　　18.5<br>• Driest Eight-Year Period of Annual Mean Natural Flow at Lee Ferry (2012–2060) Less Than 11.2 maf:<br><br>7.0　　11.2　　15.5 | Vulnerable Traces: 23%<br><br>Vulnerability Statistics:<br>• Coverage: 92%<br>• Density: 85% | Vulnerability Definition:<br>• Change in Temperature (2012–2060) Greater Than 2 Degrees Fahrenheit<br><br>−0.7　　2.0　　5.6<br>• Average Precipitation (2012–2060) Less Than 100% of Historical Average<br><br>84　　100　　111 |

NOTE: As noted in the discussion above, the percentage of vulnerable traces is somewhat higher in the Climate Conditions because only a subset of Basin Supply Scenarios (Historical and Future Climate) is used here, versus all four supply scenarios for the Streamflow Conditions.

included in the Low Historical Supply conditions.[2] Lower Basin objectives are not met in 72 percent of traces with these conditions (density). The conditions also describe 86 percent of all traces that lead to unacceptable results (coverage). And, as was true in the two Upper Basin sets of conditions, only two input parameters are needed to describe such vulnerable conditions, making them easy to understand and interpret. Figure 4.4, a scatterplot with the same dimensions as Figure 4.2, summarizes the Low Historical Supply conditions using streamflow inputs.

Recall that across all the scenarios considered in the Basin Study, there are many more traces that lead to not meeting Lower Basin objectives (47 percent) than those that lead to not meeting Upper Basin objectives (19 percent); this fact is reflected in the generally less extreme ranges included in the Lower Basin vulnerable conditions. We see substantial sensitivity with any decline from the historical mean flow coupled with an eight-year drought similar to what has been observed in the more distant past. This suggests that Lake Mead could drop below 1,000 feet even without a changing climate, if future flows resemble somewhat drier periods

**Figure 4.4**
**Low Historical Supply Vulnerable Conditions for Lake Mead Pool Elevation (Streamflow Variables)**

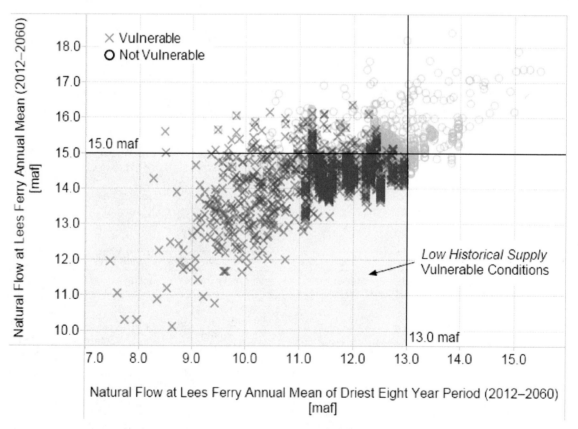

NOTE: This scatterplot shows results from thousands of individual traces, many of which overlap.
RAND RR242-4.4

---

[2]   These conditions assume that the Interim Guidelines are extended after 2026. If reservoir management instead reverts to the 2007 EIS No Action Alternative, slightly different definitions were identified using the same characterizations as streamflow. These alternate vulnerable conditions are provided in Appendix A of this report.

in the recent or paleo historical record. Once again, future demand is not included in these conditions, but growing demand remains an important factor driving growing vulnerability across all demand scenarios.

We also defined vulnerable conditions for Lower Basin delivery reliability using climate inputs to describe supply in the Historical and Future Climate supply scenarios. This investigation resulted in two sets of vulnerable conditions: one that describes key drivers when temperature is increasing, and one that describes drivers of unacceptable outcomes when temperature is consistent with the historical record.

The first part of the vulnerable conditions for Lake Mead includes a temperature cutoff of greater than a 1.7-degree Fahrenheit increase from 2012 to 2060 relative to the 1950–1999 baseline (Figure 4.5, middle vertical line). This set of conditions also includes a precipitation level of less than 104 percent of the historical average (upper horizontal line). This excludes all the traces from the Historical supply scenario (those grouped together in the left of the graphic) and some traces from the Future Climate scenario (the more dispersed traces in the center and right of the graph). In a warming climate, Lower Basin objectives are most often met only in traces that are at least 4 percent wetter than the historical average, while the large majority of traces that are drier include at least one year in which Lake Mead drops below 1,000 feet. The second part of the vulnerable conditions is insensitive to temperature and includes conditions in which precipitation declines from historical (i.e., less than 100 percent, the lower horizon-

**Figure 4.5**
**Low Historical Supply Vulnerable Conditions for Lake Mead Pool Elevation (Climate Variables)**

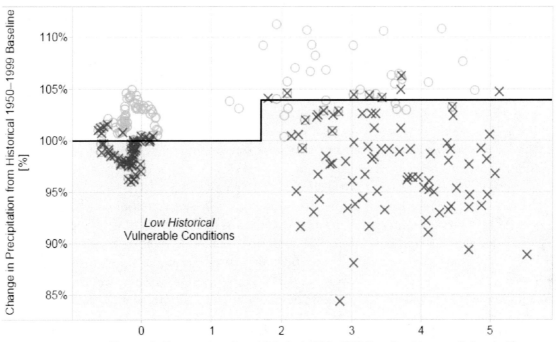

NOTES: Only showing results for the Revert to Pre-2007 No Action Alternative scenario. This scatterplot shows results from thousands of individual traces, many of which overlap. The tight cluster of results on the left correspond to the Historical supply scenario and reflect the similarity in conditions across each trace, per the ISM methodology.

RAND RR242-4.5

**Table 4.2**
**Vulnerable Conditions Defined for Lake Mead Pool Elevation: Low Historical Supply**

| Streamflow Conditions | | Climate Conditions | |
|---|---|---|---|
| Vulnerable Traces: 53%<br><br>Vulnerability Statistics:<br>• Coverage: 92%<br>• Density: 75% | Vulnerability Definition:<br>• Annual Mean Natural Flow at Lee Ferry (2012–2060) Less Than 15.0 maf per year:<br><br>10.0    15.0    18.5<br><br>• Driest Eight-Year Period of Annual Mean Natural Flow at Lee Ferry (2012–2060) Less Than 13.0 maf per year:<br><br>7.0    13.0    15.5 | Vulnerable Traces: 58%<br><br>Vulnerability Statistics:<br>• Coverage: 85%<br>• Density: 86% | Vulnerability Definition:<br>• Change in Temperature (2012–2060) Greater Than 1.7 Degrees Fahrenheit<br><br>−0.7  1.7    5.6<br>• Average Precipitation (2012–2060) Less Than 104% of Historical Average<br><br>84    104   111<br><br>OR<br>• Change in Temperature (2012–2060) Less Than 1.7 Degrees Fahrenheit<br><br>−0.7  1.7    5.6<br>• Average Precipitation (2012–2060) Less Than 100% of Historical Average<br><br>84    100   111 |

NOTE: These conditions assume that the Interim Guidelines are extended beyond 2026. As noted, the percentage of vulnerable traces is somewhat higher in the Climate Conditions because only a subset of Basin Supply Scenarios (Historical and Future Climate) is used here, versus all four supply scenarios for the streamflow conditions.

tal line). These conditions are all derived from traces from the Historical supply scenario. The results corresponding to the scenarios in which the Interim Guidelines are extended past 2026 are slightly different and shown in Appendix A.

Table 4.2 summarizes the Low Historical Supply vulnerable conditions.

Once again, a comparison of the streamflow and climate conditions identified for the Lower Basin show substantial overlap—79 percent of traces defined using climate inputs are also included in the streamflow vulnerable conditions—and we consider them to be two different ways of describing a very similar set of conditions. As a result, in subsequent chapters we will refer to these collectively as Low Historical Supply, and for convenience, we will use the streamflow-defined conditions to illustrate potential vulnerability reduction when new investment options are tested.

## Summary

In this chapter, we investigated the reliability of future water deliveries from the Colorado River using current management approaches and existing infrastructure. This analysis identified the conditions in which the current Colorado River management system would not meet its goals. These results are based on an expansive exploration of plausible future traces developed by combining supply, demand, and operations scenarios, and they make no assumptions about the likelihoods of vulnerable conditions.

Using statistical clustering methods, we defined vulnerable conditions for Upper Basin and Lower Basin Reliability—Declining Supply and Low Historical Supply, respectively. These conditions were described both in terms of streamflows and climate inputs. Declining Supply conditions include low mean streamflow (long-term mean below 13.8 maf) or Upper Basin precipitation below 100 percent of the long-term historical average. Low Historical Supply conditions, alternately, include a wider range of streamflow (long-term mean below 15 maf) or precipitation (104 percent of historical average) conditions. Conditions similar to those in the Low Historical Supply vulnerable conditions have been observed in the recent or more distant past, while those in the Declining Supply scenario reflect a shift toward lower streamflow or precipitation levels not previously observed. This suggests a strong need to evaluate what would be required to reduce these vulnerabilities.

In the next chapter, we will compare the performance of different portfolios of options in terms of meeting objectives in these conditions. We will also consider the attributes of the options needed, including cost. In Chapter Six, we will discuss how to consider likelihoods of the vulnerable conditions when considering trading off between different investment approaches.

# CHAPTER FIVE

# Reducing Vulnerabilities Through New Management Options

The vulnerabilities identified and described in Chapter Four represent significant threats to the successful management of the Colorado River. The analysis shows that if no changes are made—that is, if the current management approach continues—there are many plausible futures in which Upper and Lower Basin objectives would not be met.

The Basin Study developed four different portfolios and evaluated how they could improve outcomes. The portfolios, discussed in Chapter Three, were defined to be dynamic. This means that the simulation model is programmed to implement options only if the simulated Basin conditions warrant them. Taking advantage of this feature, CRSS calculated a unique sequence of options to be implemented for each of the 23,508 traces comprising the demand, supply, and reservoir options scenarios. In this chapter, we present the effects of the portfolios on improving performance of the system across all future traces as well in the vulnerable conditions identified in Chapter Four. We then examine the frequency with which different options are implemented for each dynamic portfolio. We end by exploring the trade-offs among the different portfolios in terms of their effectiveness at reducing vulnerabilities and the cost of their implementation.

## How Well Do Portfolios of Options Reduce Vulnerabilities?

We evaluated four portfolios—*Portfolio A (Inclusive), Portfolio B (Reliability Focus), Portfolio C (Environmental Performance Focus)*, and *Portfolio D (Common Options)*—across all the scenarios. We then reviewed how each performed under the vulnerable conditions described in Chapter Four—Declining Supplies for the Upper Basin and Low Historical Supplies. As expected, implementation of the portfolios reduces the number of years and traces in which the system fails to meet Basin goals across many scenarios.

This section reviews the effects of the portfolios on vulnerability reduction for the Upper Basin Reliability metric (Lee Ferry Deficit) and the Lower Basin Reliability metric (Lake Mead Pool Elevations). Figure 5.1 and Figure 5.2 show results in terms of reductions in the number of traces and number of years in which the Upper Basin and Lower Basin goals are not met. We also group the results by all traces, and just for those traces corresponding to the Low Historical Supply and Declining Supply vulnerable conditions.

For Upper Basin Reliability, Figure 5.1 shows that implementation of the portfolios reduces the percentage of years in which deficits occur to between 1 and 2 percent of all traces from a baseline of 4 percent (top row, right column), and reduces the number of traces with a single Upper Basin Reliability vulnerability to between 6 and 13 percent from a baseline of

**Figure 5.1**
**Reduction in Upper Basin Reliability Vulnerability (Lee Ferry Deficit) Across Portfolios**

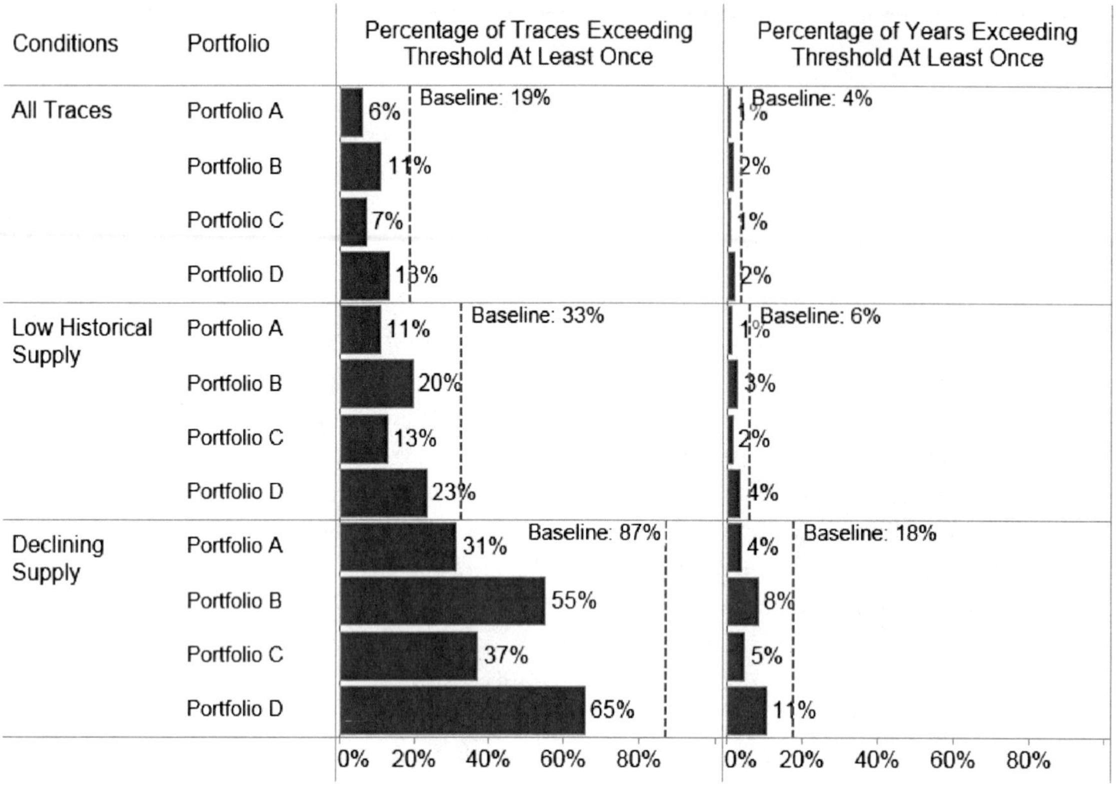

| Conditions | Portfolio | Percentage of Traces Exceeding Threshold At Least Once | Percentage of Years Exceeding Threshold At Least Once |
|---|---|---|---|
| All Traces | Portfolio A | 6% — Baseline: 19% | 1% — Baseline: 4% |
| | Portfolio B | 11% | 2% |
| | Portfolio C | 7% | 1% |
| | Portfolio D | 13% | 2% |
| Low Historical Supply | Portfolio A | 11% — Baseline: 33% | 1% — Baseline: 6% |
| | Portfolio B | 20% | 3% |
| | Portfolio C | 13% | 2% |
| | Portfolio D | 23% | 4% |
| Declining Supply | Portfolio A | 31% — Baseline: 87% | 4% — Baseline: 18% |
| | Portfolio B | 55% | 8% |
| | Portfolio C | 37% | 5% |
| | Portfolio D | 65% | 11% |

NOTE: Figure is expressed in terms of percentage of traces (left) and percentage of years (right) across all traces (top row) and Low Historical Supply and Declining Supply vulnerability conditions (bottom two rows).
RAND RR242-5.1

19 percent (left column). The differences between the portfolios in terms of addressing Upper Basin vulnerabilities can be seen most clearly when focusing on the Declining Supply vulnerable conditions. *Portfolio C (Environmental Performance Focus)* reduces vulnerabilities in 18 percent more traces (37 percent) than *Portfolio B (Reliability Focus)*, at 55 percent.

Figure 5.2 shows the same results for the Lower Basin Reliability metric: Lake Mead Pool Elevation. Similar to the Upper Basin results, the implementation of the portfolios significantly reduces the number of years in which the Basin goals are not met. Even in the most stressing Declining Supply vulnerable conditions, the percentage of years is reduced from 50 percent to between 22 and 27 percent (right side of Figure 5.2). Unfortunately, these reductions in yearly vulnerability do not lead to significantly fewer traces in which Lake Mead elevation drops below 1,000 feet in at least one year (left side of Figure 5.2). The results do suggest that *Portfolio C (Environmental Performance Focus)* is more effective at reducing Lower Basin vulnerability than the *Reliability Focus* portfolio.

The implementation of portfolios increases the robustness of the system and shrinks the set of conditions in which the system does not meet its goals. For example, with *Portfolio A (Inclusive)*, the conditions in which the Upper Basin goals are not met are reduced. The vulnerable conditions in terms of mean natural flow at Lees Ferry declines from less than 15.0 maf per year to less than 13.2 maf per year, and the flow during the eight-year dry period declines from

**Figure 5.2**
**Reduction in Lower Basin Vulnerability (Lake Mead Pool Elevations) Across Portfolios**

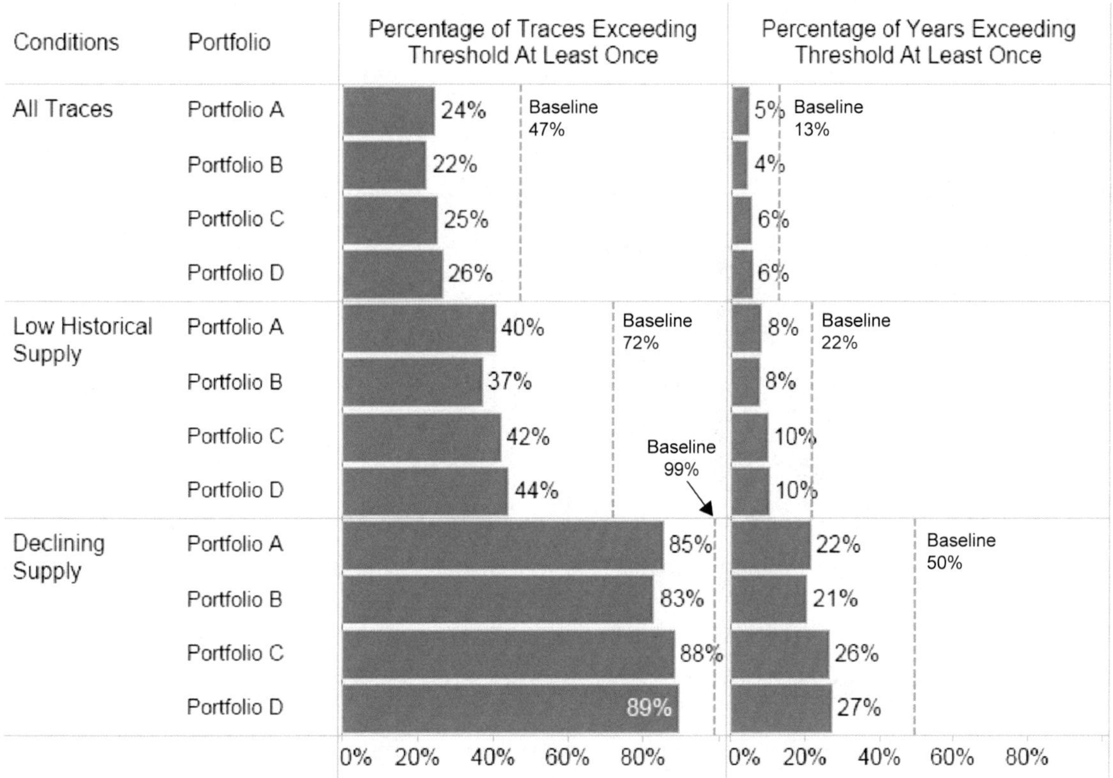

| Conditions | Portfolio | Percentage of Traces Exceeding Threshold At Least Once | Percentage of Years Exceeding Threshold At Least Once |
|---|---|---|---|
| All Traces | Portfolio A | 24% | Baseline 47% | 5% | Baseline 13% |
| | Portfolio B | 22% | 4% |
| | Portfolio C | 25% | 6% |
| | Portfolio D | 26% | 6% |
| Low Historical Supply | Portfolio A | 40% | Baseline 72% | 8% | Baseline 22% |
| | Portfolio B | 37% | 8% |
| | Portfolio C | 42% | 10% |
| | Portfolio D | 44% | Baseline 99% | 10% |
| Declining Supply | Portfolio A | 85% | 22% | Baseline 50% |
| | Portfolio B | 83% | 21% |
| | Portfolio C | 88% | 26% |
| | Portfolio D | 89% | 27% |

0%  20%  40%  60%  80%        0%  20%  40%  60%  80%

NOTE: Figure is expressed in terms of percentage of traces (left) and percentage of years (right) across all traces (top row) and Low Historical Supply and Declining Supply vulnerability conditions (bottom two rows).
RAND RR242-5.2

less than 11.2 maf per year to less than 10.0 maf per year (Figure 5.3, top panel). In terms of the climate variables, the new vulnerable conditions with *Portfolio A (Inclusive)* in place increases the temperatures more than 2.6 degrees Fahrenheit and lowers the precipitation from less than 100 percent of the historical Upper Basin average to less than 96 percent of the historical average. This represents a significant reduction of the range of conditions that produce Upper Basin vulnerability. In the figure, the dashed lines and green shading correspond to the original Declining Supply vulnerable conditions. The solid lines and yellow region correspond to the reduced Declining Supply vulnerable conditions because of the implementation of *Portfolio A (Inclusive)*.

A similar, though not identical, story emerges for the reduction in vulnerability for the Lower Basin with *Portfolio A (Inclusive)*. This portfolio shrinks the range of streamflow conditions in which Lake Mead level does not meet the Basin's goals: The long-term mean streamflow conditions decline from less than 15.0 maf to less than 14.5 maf, and eight-year drought streamflow conditions change from less than 13 maf to less than 12 maf (Figure 5.4, top panel). This translates to a nearly complete elimination of unmet goals for climate conditions similar to historical conditions (Figure 5.4, bottom panel), while further restricting the temperature and precipitation range from the climate scenario in which Lower Basin goals are not met. In the figure, the dashed lines and green shading correspond to the original Low Historical Supply vulnerable

**Figure 5.3**
**Change in Streamflow Conditions and Climate Conditions Leading to Upper Basin Vulnerability for** *Portfolio A (Inclusive)*

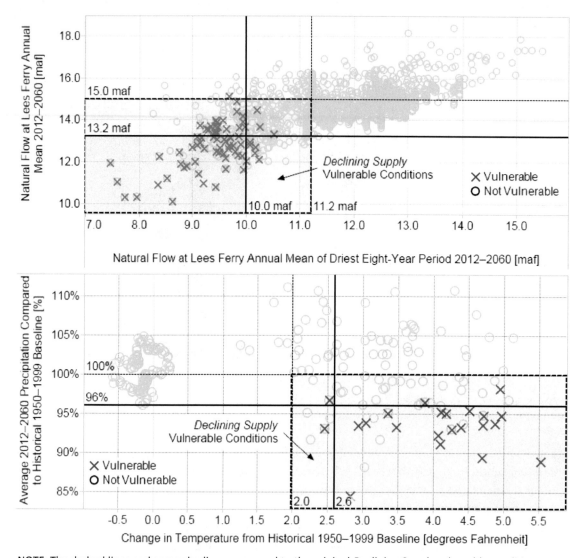

NOTE: The dashed lines and green shading correspond to the original *Declining Supply* vulnerable conditions. The solid lines and yellow region correspond to the reduced vulnerable conditions due to the implementation of *Portfolio A (Inclusive)*. This scatterplot shows results from thousands of individual traces, many of which overlap.
RAND RR242-5.3

conditions. The solid lines and yellow region correspond to the reduced Low Historical Supply vulnerable conditions because of the implementation of *Portfolio A (Inclusive)*.

## Which Options Are Most Needed to Address Emerging Vulnerabilities?

Each CRSS simulation of a dynamic portfolio defines the sequences in which options are implemented. When viewed across all traces, one can see which options are implemented and when they are implemented relative to the time they become available.

**Figure 5.4**
**Change in Streamflow Conditions and Climate Conditions that Lead to Lower Basin Vulnerability for**
***Portfolio A (Inclusive)***

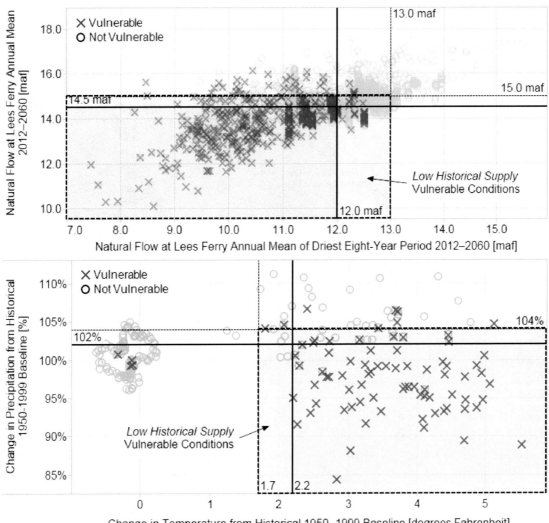

NOTE: The dashed lines and green shading correspond to the original Low Historical Supply vulnerable conditions. The solid lines and yellow region correspond to the reduced vulnerable conditions due to the implementation of Portfolio A (Inclusive). This scatterplot shows results from thousands of individual traces, many of which overlap.
RAND RR242-5.4

Using a few representative examples from *Portfolio D (Common Options)*, Figure 5.5 shows the frequency of implemented options. Table 5.1 provides a brief description of the option types discussed in this chapter, and Appendix B provides extended descriptions. For example, the Agricultural Conservation LB-Transfer (Step 1) option is available in 2016. By 2019, it is implemented in 25 percent of all traces evaluated. By 2028, it is implemented in more than 80 percent of traces. The Reuse-Industrial option, in contrast, is available in 2021 and is not implemented in more than 20 percent of traces until 2039 (18 years after it is available). By 2060, it is only implemented in about 40 percent of the traces. Note that the fifth step of M&I Conservation is implemented in about 40 percent of traces as soon as it is available. This sug-

**Figure 5.5**
**Percentage of Traces in Which Five Options Are Implemented by Year for *Portfolio D (Common Options)***

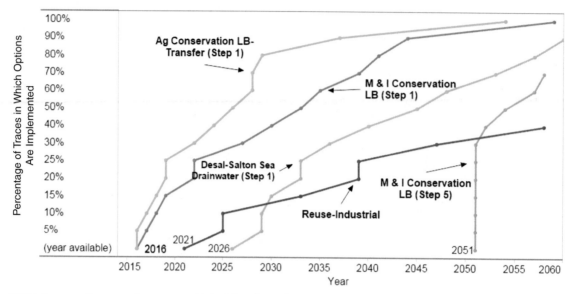

NOTE: Large options are disaggregated into 200 taf/year "steps" to represent likely project phasing.
RAND RR242-5.5

gests that in many plausible futures, there is a need for substantial additional conservation before it is assumed to be attainable in the simulation.

Extending these results to all options in *Portfolio D (Common Options)*, Figure 5.6 shows by year (horizontal axis) the percentage of traces in which each of the options are implemented (color of the symbol). Symbols colored blue indicate options implemented in very few traces, while those colored orange or red are implemented in most or all traces. This figure shows that, early in the simulation period, the Agricultural and M&I Conservation options are implemented most frequently. By the end of the period, other options implemented in most traces include Desal-Yuma Area Groundwater and Desal-Salton Sea Drainwater. Notably, almost all options are implemented in at least some traces soon after they are available, indicating a compelling need to reduce vulnerabilities that have already emerged in these traces.

## What Are the Key Trade-Offs Among Portfolios?

The four portfolios span a range of investment strategies for addressing vulnerabilities, each with unique costs and effects on vulnerabilities. Figure 5.7 shows the distribution of costs across the simulations by year for each of the four portfolios (columns) and for three sets of traces—all traces, those in the Low Historical Supply vulnerable conditions, and those in the Declining Supply vulnerable conditions. Implementation costs increase over time as options are implemented in response to the signposts, and there is a wide range in costs across the traces. For *Portfolio A (Inclusive)*, the costs range from just under $2 billion per year to more than $7 billion per year in 2060. This wide range of costs indicates that the dynamic portfolios

**Table 5.1**
**Option Types Included in the Portfolio Analysis**

| Option Type | Description | Options Included in *Portfolio A (Inclusive)* |
|---|---|---|
| Agricultural Conservation | Options that increase water conservation in the agriculture sector and reduce demand for Colorado River water in either the Upper or Lower Basin: Options are disaggregated into 200 taf per year "steps" to represent likely project phasing | Agricultural Conservation with Transfers (Upper and Lower Basin) |
| Desalination | Options to desalinate (1) ocean water off the California and the Gulf of Mexico coasts, (2) agricultural drainwater, and (3) brackish groundwater | Desal–Salton Sea Drainwater Desal–Pacific Ocean–California Desal-Gulf |
| Energy Water Use Efficiency | Options to improve the water use efficiency of the energy sector | Energy Water Use Efficiency– Air Cooling |
| Import | Options to increase the overall water supply of the Basin from other river basins: Imports from the Missouri River and The Mississippi River were considered to augment supply in the Colorado Front Range and reduce the amount of Colorado River exported to these regions. | Import–Front Range–Missouri |
| Local Supply | Local supply options capture local water sources that would otherwise go unused | Local-Coalbed Methane |
| M&I Conservation | Options that increase water conservation in the municipal and industrial sectors and reduces demand for Colorado River water in either the Upper or Lower Basins | M&I Conservation (Upper and Lower Basin) |
| Reuse | Reuse of existing municipal and gray water supplies increase the overall water supply in the Basin | Reuse-Municipal Reuse–Gray Water |
| Watershed Management | Options that could increase the supply of the Basin by increasing river runoff: Key approaches include Tamarisk control, Forest Management, Brush Control, Dust Control, and Weather Modification | Watershed–Weather Mod Watershed-Dust |
| Groundwater Banking | Option creates an Upper Basin water bank in either Lake Powell or in an offstream groundwater bank to increase protection against curtailment in the Upper Basin | Upper Basin Groundwater Bank |

NOTE: Large options are disaggregated into 200 taf/year "steps" to represent likely project phasing.

as designed for the study help restrain unnecessary investment when conditions do not warrant it.

However, when looking at just the traces in the Declining Supply vulnerable conditions (bottom row), the range in costs is narrower. By 2060, costs for most traces are close to the maximum observed cost. This suggests that under the Declining Supply conditions, all available options are implemented by the end of the simulation period. In contrast, traces within the Low Historical Supply conditions span a wide range in costs, suggesting that in many cases, more modest investments are required to eliminate vulnerabilities.

Lastly, looking across the portfolios, one sees that the cost range of *Portfolio A (Inclusive)* is generally higher than the other three portfolios by the end of the 49-year period. This is due, in part, to the inclusion of more expensive options than the other three portfolios coupled with the need to implement as many options as are available in the portfolio to address vulnerabilities under many traces. This further suggests that the other three portfolios are more limited by their more restricted set of available options.

**Figure 5.6**
**Frequency of Option Implementation (percentage of traces) for *Portfolio D (Common Options)***

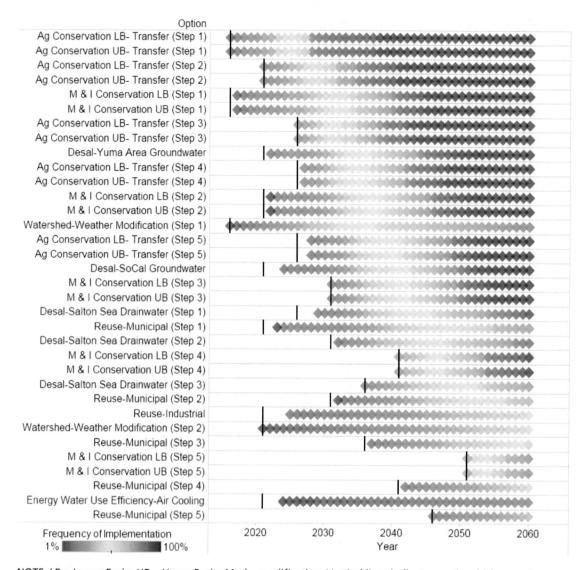

NOTE: LB = Lower Basin; UB = Upper Basin; Mod = modification. Vertical lines indicate year in which an option could be available for implementation.
**RAND** *RR242-5.6*

We can now combine the cost and vulnerability results together to draw out the distinctions and trade-offs among the four portfolios. First, there is little difference between portfolios when looking across all traces evaluated. That is, the range in vulnerability reduction and costs overlap significantly for all the portfolios (Figure 5.8, top panel). This is not surprising, because there are many traces evaluated in which there is only a modest need for improvement and all four of the portfolios can address those needs using options with similar costs.

When we focus on traces corresponding to the vulnerable conditions, we see some modest difference across the portfolios. First, in the Low Historical Supply conditions (Figure 5.8, middle panel), we see that the portfolios with more options reduce the number of years in which the Upper Basin and Lower Basin goals are not met. The ranges in costs (horizontal

**Figure 5.7**
**Distribution of Total Annual Cost Resulting from Implementation of the Dynamic Portfolios for All Traces, Low Historical Supply, and Declining Supply**

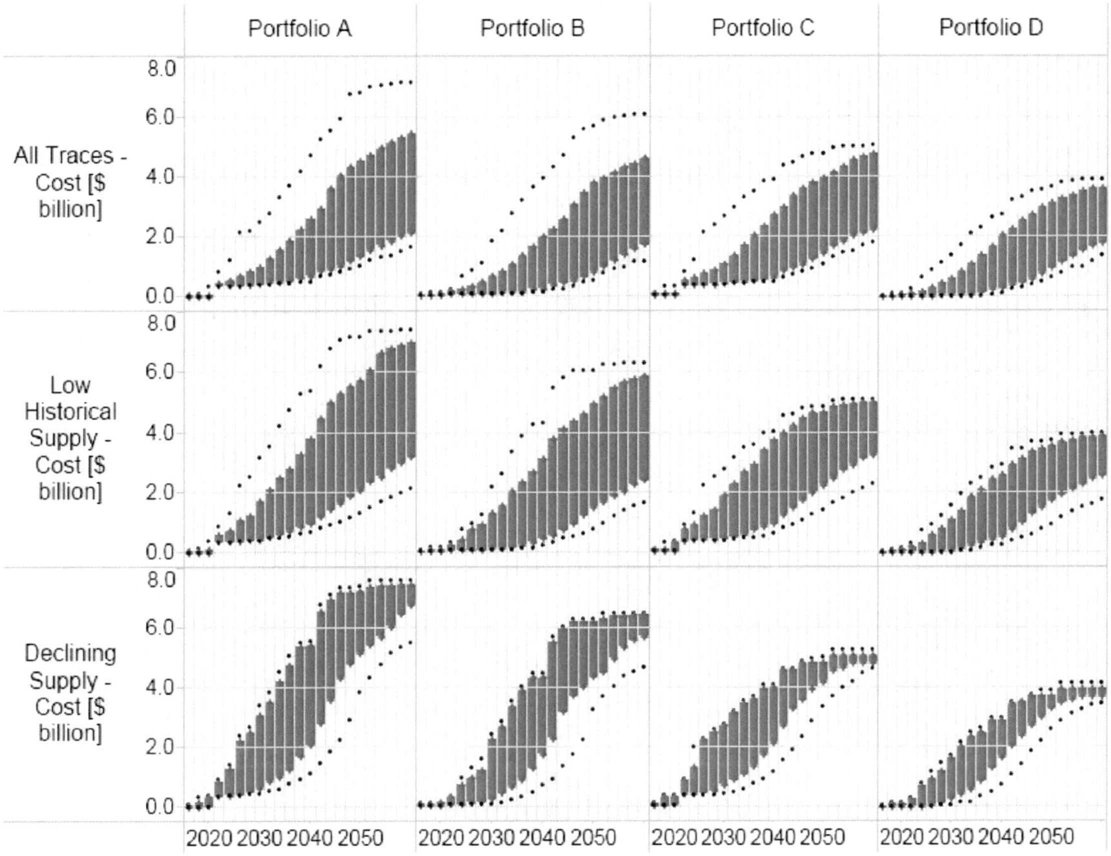

spread) across the traces widens significantly, but there is again significant overlap among the portfolios. When we only include traces in the Declining Supply vulnerable conditions, the trade-offs become clear. For the Upper Basin, *Portfolio C (Environmental Performance Focus)* is not only more effective than *Portfolio B (Reliability Focus)* and *Portfolio D (Common Options)*, it costs significantly less than *Portfolio B (Reliability Focus)*. Only *Portfolio A (Inclusive)* reduces vulnerability more, but at significantly higher cost. *Portfolio C (Environmental Performance Focus)* dominates because it includes an Upper Basin water bank, which is used to maintain flow to the Lower Basin at Lee Ferry, and excludes other, more expensive, new supply options.

However, performance with respect to the Lower Basin objectives (Figure 5.8, right panel) shows that *Portfolio B (Reliability Focus)* improves reliability as well as or better than the other portfolios in all three sets of conditions. *Portfolio B (Reliability Focus)* includes more options that directly benefit the Lower Basin, including Pacific Ocean desalination projects. Given this more-focused investment, *Portfolio B (Reliability Focus)* dominates *Portfolio A (Inclusive)* by being just as effective but less costly.

**Figure 5.8**
**Trade-Offs Between Portfolio Costs and Vulnerabilities Across Portfolios**

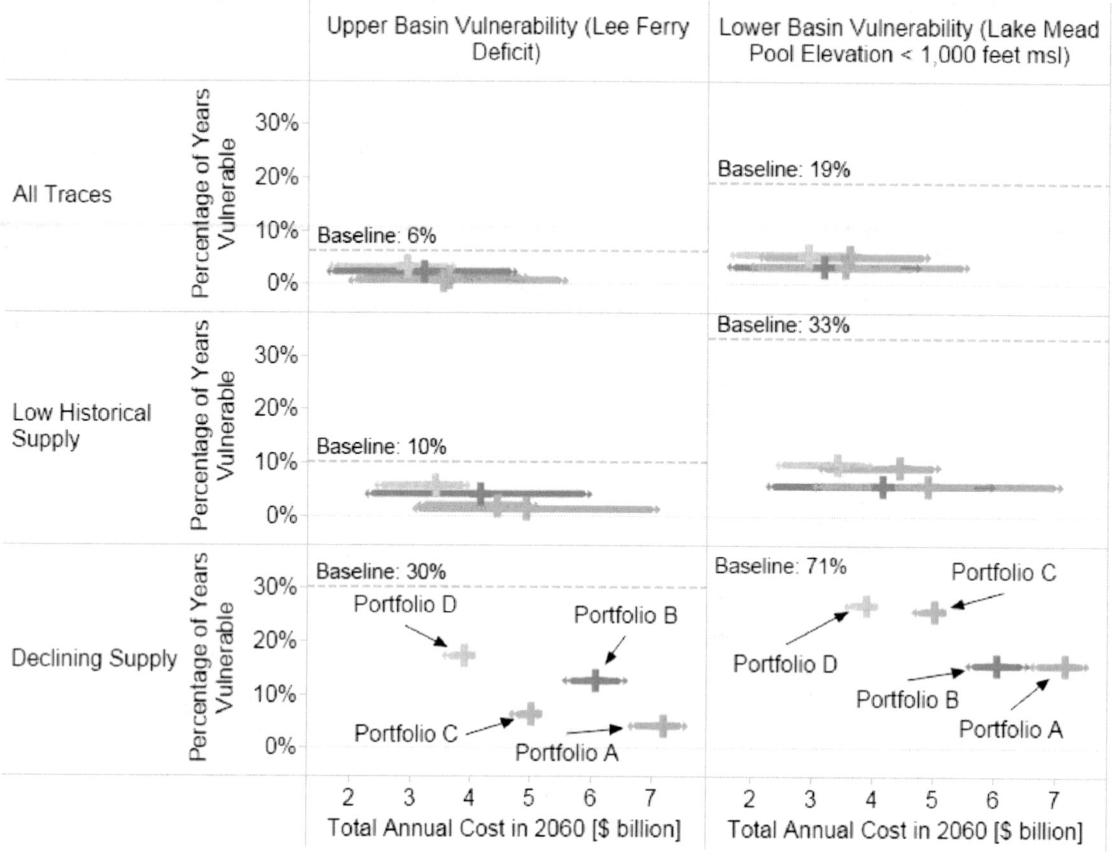

## Summary

In this chapter, we described how four portfolios including different sets of supply-augmentation or demand-reduction investments could increase the reliability of future Colorado Basin deliveries. Our analysis shows that some system vulnerability described in Chapter Four could be eliminated by implementing a wide array of new options. We explored which options would be implemented for each portfolio in order to identify those options most frequently implemented across many or most traces. Next, we reviewed the costs incurred when implementing options included in different portfolios. Costs could be high when vulnerable conditions emerge, with *Portfolio A (Inclusive)* costs exceeding $7 billion per year by 2060, for example, under the Declining Supply vulnerable conditions. Lastly, we identified several important trade-offs among the portfolios. Specifically, *Portfolio C (Environmental Performance Focus)* best addresses Upper Basin Reliability vulnerabilities at the lowest comparative cost, while *Portfolio B (Reliability Focus)* performs similarly well in terms of performance and cost for Lower Basin Reliability.

None of the portfolios, however, eliminate the possibility of not meeting the Basin's water delivery objectives across all future conditions we tested. For all portfolios, Lee Ferry Deficits could occur in 4 percent of the years across the traces in the Declining Supply vulnerable conditions. In 15 percent of the years, Lake Mead elevations could also fall below 1,000 feet. When considering the entire traces, more modest reductions are seen. Even with the implementation of *Portfolio A (Inclusive)*, Lee Ferry Deficits occur in 11 percent of the traces included in the Low Historical Supply vulnerable conditions and in 31 percent of the traces included in the Declining Supply vulnerable conditions.

A few words of caution about these results are warranted. First, the four portfolios evaluated represent only a few of many possible portfolios that different stakeholders and decision makers might want to consider. Though suggestive, the current portfolio analysis cannot be viewed as sufficient to choose which investments to make and when. Second, the portfolios were modeled such that an option could be implemented as long as the simulated time period is later than the earliest time in which an option is available. If an option is to be implemented very soon after it is available in some futures, that implies that feasibility, planning, and construction would need to begin soon—even if the option is not ultimately implemented in other futures. This requires the careful consideration of which options the Basin will begin to implement now and which to delay.

In the next chapter, we describe how the analysis can help inform the implementation of a robust and adaptive Colorado River Basin investment strategy over the next 50 years.

# CHAPTER SIX

# Implementing a Robust, Adaptive Strategy for the Colorado River Basin

The Basin Study examined how a wide range of water-management options, when implemented as part of different dynamic portfolios, could reduce the Colorado River Basin's vulnerabilities to changing and uncertain future conditions. It does not, however, recommend a specific management strategy. Instead, the Basin Study presents key trade-offs among portfolios in terms of costs and vulnerability reduction, and highlights which options are most needed and when. The report concludes by describing a range of near-term steps to be undertaken to move toward development and implementation of a broader investment strategy for the Basin.

As described in Chapter Five, a predefined or static strategy of management options is unlikely to meet Basin goals cost-effectively, as different plausible future conditions will require different responses. Instead, the Basin Study sought to evaluate strategies that are adaptive over time in order to improve robustness to challenging future conditions while avoiding excess costs when conditions are more benign. The analysis described in this report provides important information needed to develop such a robust, adaptive strategy. Specifically, this analysis helped develop and test four initial adaptive strategies that all include a set of near-term options, a procedure for monitoring evolving conditions, and a set of additional options to be taken if conditions warrant. Below, we discuss how the Basin Study analysis can be used to build on each of these elements in future planning efforts.

## Near-Term Options as the Foundation of a Robust Strategy

As described in Chapter Five, for each portfolio we identified those options that are almost always needed regardless of differing assumptions about future conditions. In particular, because Portfolio D (Common Options) includes only options selected for both of the two stakeholder-derived portfolios (Portfolios B and C), options always or frequently implemented in this portfolio as soon as they are available can be considered both near-term and high priority.

Figure 6.1 summarizes how frequently options from Portfolio D (Common Options) are implemented by 2060 (horizontal axis) and the delay in their implementation (vertical axis), expressed as the median delay across all traces relative to the time they become available. The results are presented for three sets of traces—all traces (top panel), those traces in the Low Historical Supply vulnerable conditions (middle panel), and those traces in the Declining Supply vulnerable conditions (bottom panel).

Results in the lower-right corner of the all traces panel (top) are near-term, high-priority options. Specifically, M&I Conservation is shown to be required in more than 90 percent of all traces examined in the study with a minimum delay of only one year. Agricultural Conservation with Transfers is implemented in almost 100 percent of traces, but with a delay of six years. Three more desalination options—Desal-Salton Sea, Desal-Yuma, and, Desal-Groundwater—are all high-priority but are needed only after delays of eight years or more.

For future conditions consistent with the two key vulnerable conditions—Low Historical Supply and Declining Supply—more options are needed, and with less delay. The middle panel of Figure 6.1 shows that for the Low Historical Supply vulnerable conditions, the urgency of implementation of Agricultural Conservation with Transfers and Desal-Salton Sea increases, making them both near-term, high-priority options. The Reuse-Municipal option is also required in more than 70 percent of traces. The bottom panel shows that for Declining Supply vulnerable conditions, all options in Portfolio D (Common Options) are needed by 2060 in nearly all traces.

Figure 6.1 shows that most of the options in the Portfolio D (Common Options) are needed in only some future traces, and in many cases are implemented only after a delay. However, the conditions corresponding to the Low Historical vulnerable conditions have been experienced in the recent past and those corresponding to the Declining Supply are predicted by many global climate model simulations. As the Basin Study highlights, the Basin needs not commit to all possible options now, but it must use the available lead time to prepare to invest in new options if conditions suggest they are warranted. The implementation of some options with longer lead times will need to be initiated soon so they will be available if needed under particular future traces. **Exploring plans during this time for design and permitting of selected options would provide decision makers with a hedge against potential delays in implementation if the options are needed in response to changing conditions.**

## Monitoring Conditions to Signal Implementation of Additional Options

The Basin Study used relatively simple streamflow and storage condition signposts, programmed into the CRSS model, to identify when options would need to be implemented to significantly reduce vulnerabilities. As shown in Table 3.8 in Chapter Three, many of the signposts are based on observing a combination of low streamflow conditions and low elevations for Lakes Powell and Mead. The precursor conditions to natural streamflow are primarily climate parameters, including precipitation and temperature. Reclamation and other agencies are already collecting critical information (e.g., streamflow, climate conditions, status of the reservoirs) that can be used to inform assessments of which options should be implemented in the future. **Building this information into systematic and recurring system assessments would enable managers and users of the Basin to better understand how conditions are evolving and plan for additional management options accordingly.**

The vulnerability analysis specifically showed that the Upper Basin is vulnerable to climate conditions that are consistent with many of the simulated conditions emerging from a variety of global climate models. Over the next few years, it may be easier to discern whether the future climate is going continue to deviate from the historical record, drawing from new climate models or higher resolution regional climate projections (e.g., Seager et al. [2012]). In particular, additional advances in regional climate modeling could improve projections of future precipitation and runoff in the Upper Colorado Basin. **If the results from improved models are consistent with the more pessimistic current projections, the Basin is increasingly likely to face vulnerable conditions for the Lee Ferry Deficit and Lake Mead levels.**

**Figure 6.1**
**Percentage of Traces in Which Options Are Implemented and Associated Implementation Delay for**
***Portfolio D***

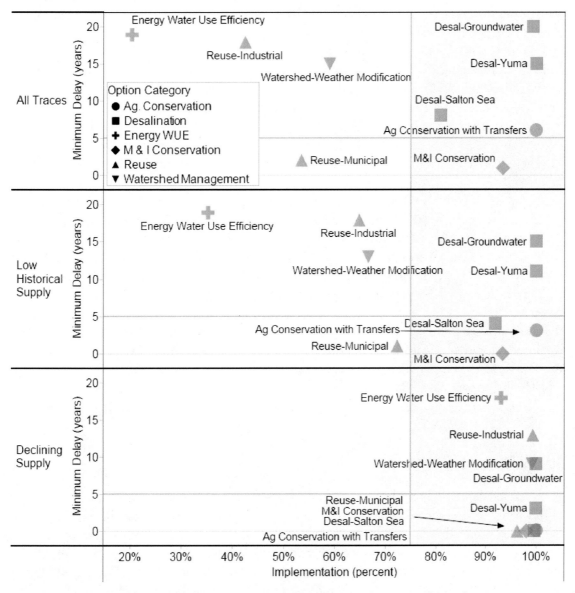

NOTE: Delay is expressed as the minimum delay across all the implemented options within each category. Implementation is averaged across all traces and options.

RAND *RR242-6.1*

**Many of the options identified as necessary under these conditions would need to be considered for implementation.**

## Options to Implement If Future Conditions Warrant

The analysis has shown that as vulnerable conditions develop in the Basin, increasingly expensive adaptation options will be required. The analysis highlighted which options would be needed and when. However, preparation for many of these options would need to begin well before their implementation. **For this mid- to longer-term implementation period of a robust, adaptive strategy, Reclamation and the Basin States could identify the key long lead-time options that may be needed and begin to take near-term planning and design steps to ensure their availability.**

**It may also be beneficial to consider additional approaches for managing future imbalances that rely less on large-scale supply augmentation and efficiency-based conservation.** Results in Chapter Five indicate how many plausible future conditions under which both Upper Basin and Lower Basin Reliability objectives will not be met. For the Upper Basin, if conditions turn out to be consistent with the hotter and drier Future Climate scenario traces, even extensive investment in the Basin still may not prevent a Lee Ferry Deficit (Figure 5.3 in Chapter Five). Remaining vulnerabilities are even more significant for the Lower Basin. Even with the significant investments identified in the Basin Study but without the warming predicted to accompany climate change, a repeat of extended droughts or long-term low flows observed in the historical record would lead to Lake Mead falling below 1,000 feet.

Throughout the course of the Basin Study, options for reconsidering the allocation and management of the Colorado were raised. Many additional adaptation options, such as some types of water transfers, are consistent with the current Law of the River, but could not be easily modeled by CRSS within the time available to complete the study. As suggested by the Basin Study, evaluating these additional options in the coming months could further improve the ability for the portfolios to address supply and demand imbalances. **Revisiting the options included in the portfolio is fully consistent with the RDM analysis framework used in the Basin Study. Comparing and contrasting the performance and other attributes of additional approaches alongside the adaptive options evaluated for the Basin Study would support the successful implementation of a robust, adaptive strategy.**

## Summary

This chapter describes how the analytic results developed for the Basin Study and highlighted in this report can support implementation of a robust, adaptive strategy for the Colorado River Basin. Specifically, the simulation of the dynamic portfolios identified a small set of options that are near-term and high-priority. We further show which options are often implemented when focusing only on the two sets of vulnerable conditions described in Chapter Three, Low Historical Supply and Declining Supply, and how these frequently implemented options differ across the portfolios. Taken together, these results can inform stakeholder and decision-maker deliberations about how aggressively to prepare for key vulnerabilities. Planners across the Basin are already beginning to develop the needed workgroups to move forward on these options.

The analysis also demonstrated that signposts related to observable climate, streamflow, and system conditions (e.g., reservoir surface elevations), could form the basis of a monitoring program and trigger the implementation of additional management options. We conclude the chapter by discussing additional analysis that could help to identify other options to further reduce vulnerabilities of the Basin. The RDM method described in this report can be applied to structure the assimilation of such new information as part of the adaptive management approach.

# Vulnerability Analysis Example and Additional Results

## Vulnerability Analysis Example

Vulnerability analysis begins with the database of simulation model results (or cases) generated in Step 2 of the RDM analysis. Analysts first define minimally acceptable outcomes or satisficing thresholds for one or more performance metrics.

For example, a useful threshold for Lake Mead Pool Elevations is 1,000 feet—the level below which the Southern Nevada Water Authority can no longer withdraw water using its lowest current intake. Figure A.1 illustrates two example simulations of Lake Mead Pool Elevation over time. If the reservoir elevation falls below the 1,000-foot threshold, the Colorado River management system does not meet its baseline water delivery objectives.

In the vulnerability analysis, the analyst next uses algorithms such as the Patient Rule Induction Method (PRIM) (Friedman and Fisher, 1999) to identify the uncertain external conditions that lead the system to not meet key objectives. The uncertain external conditions are statistical or numerical characterizations of the futures evaluated in Step 2. In water-management studies, these can include aspects of the future climate, e.g., mean temperature, precipitation (Groves and Lempert, 2007).

**Figure A.1**
**Example Simulations in Which Lake Mead Pool Elevation Objectives Are Met and Not Met**

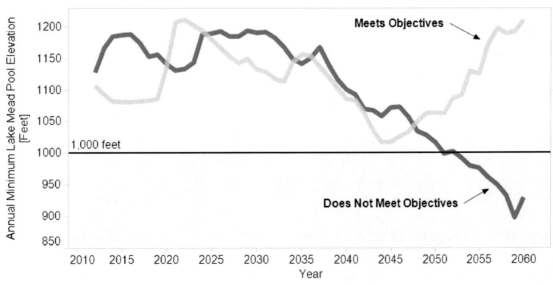

RAND *RR242-A.1*

Three measures of merit help guide this process:

- **Coverage:** the fraction of all traces where objectives are not met in the database that are contained within the vulnerable condition. Ideally, the vulnerable conditions would contain all such cases in the database and coverage would be 100 percent.
- **Density:** the fraction of all the cases within the vulnerable conditions that do not meet objectives. Ideally, all the cases within the vulnerable conditions would be vulnerable and density would be 100 percent.
- **Interpretability:** the ease with which users can understand the information conveyed by the vulnerable condition. The number of uncertain conditions used to define the scenario serves as a proxy for interpretability. The smaller the number of parameters used to describe the vulnerable conditions, the higher the interpretability.

These three measures are generally in tension with one another. For instance, increasing density may decrease coverage and interpretability. PRIM thus generates a set of vulnerable conditions and allows the decision maker and analyst to choose the one with the combination of density, coverage, and interpretability most suitable for their application. Other algorithms such as Classification and Regression Tree or principal component analysis have also been used; for analyses with a small number of uncertain factors, manual inspection can be used.

To illustrate how PRIM works, consider a set of 20 simulations, in which goals are not met in four of them. Each simulation is defined by inputs X, Y, and Z, each of which can take a value between 0 and 1. Figure A.2 shows graphically each simulation result with respect to inputs X (horizontal axis), Y (vertical axis), and Z (size of symbol). The red symbols are those that do not meet the goals and the gray symbols are those that do. The red lines and yellow shading indicate the input range restrictions for one set of vulnerable conditions. The "X" symbols are those explained by the vulnerable conditions and the "O" symbols are those not explained by the vulnerable conditions.

PRIM can define different vulnerable conditions for these results. Table A.1 lists the definition, coverage, and density results for four different sets of vulnerable conditions. The conditions shown in Figure A.2, for example, restrict the X and Y variables and describe three of the four vulnerable simulations (75 percent coverage)—and three of the four simulations described by the definition are vulnerable (75 percent density). The other vulnerability definitions shown in Table A.1 show the trade-off between coverage, density, and number of restrictions. PRIM presents a set of vulnerable conditions on the frontier of the density coverage trade-off, and the decision maker and analyst then select the set of conditions most appropriate for the decision at hand.

To illustrate these concepts using simulations of the Colorado River Basin, the top panel of Figure A.3 expands on Figure A.1 and shows ten simulation results of Lake Mead Pool Elevation for ten plausible future traces. Using a more complete set of simulations, PRIM defined a set of external conditions that generally lead to outcomes in which Lake Mead drops below 1,000 feet. Figure A.3 classifies the ten simulations as not being consistent with the vulnerable conditions and also meeting Lake Mead goals (thin gray lines), being consistent with the vulnerable conditions and not meeting Lake Mead goals (red lines), and being consistent with the vulnerable conditions but meeting Lake Mead goals (yellow-green lines). The goal of scenario discovery is to define vulnerable conditions that explain all the cases in which goals are not met (the red lines in Figure A.3) but none of the cases in which goals are met (the yellow-green lines in Figure A.3).

**Figure A.2**
**20 Simulation Outcomes with Definitions of Vulnerable Conditions**

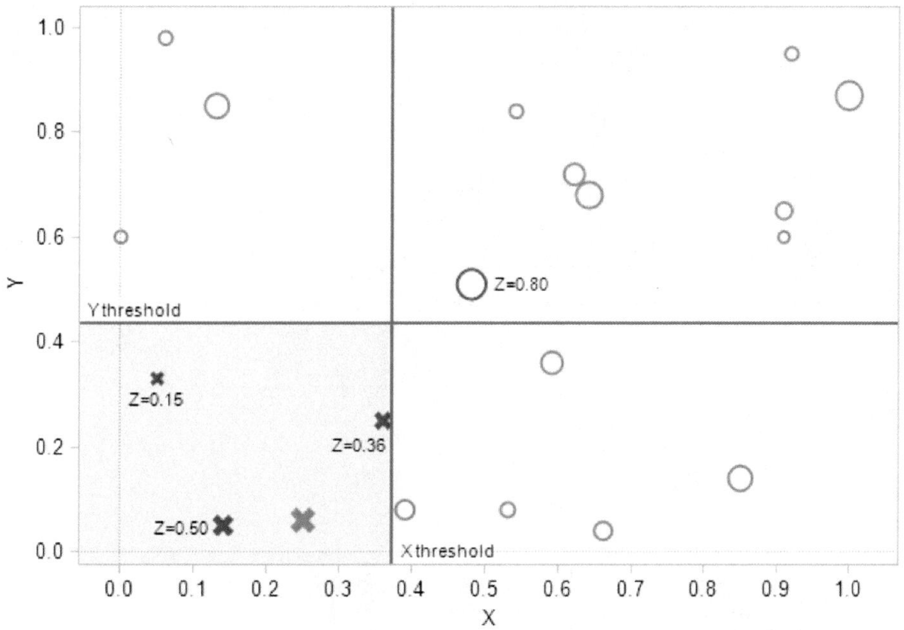

NOTE: Red symbols indicate simulations that are vulnerable. "X" symbols indicate simulations described by the Vulnerable Conditions. X and Y boundaries are shown by the red lines. The size of each point is defined by the Z variable. The yellow-shaded region represents the region of the input space that defines the vulnerability. There is no restriction on the Z variable for this vulnerability description.

RAND RR242-A.2

## Lake Mead Vulnerable Conditions Without Interim Guidelines Post-2026

Another set of conditions focuses on traces in which the trend in increased temperatures is no greater than 1 degree Fahrenheit. Here, demand emerges as a key driver of vulnerability for the first time. If demand between 2041 and 2060 is greater than 14.5 maf per year, average precipitation falls below 100 percent of the 1950–1999 average, and the reservoir operations

**Table A.1**
**Example Statistics for Several Different Example PRIM Vulnerable Conditions**

| Definition | Coverage | Density |
|---|---|---|
| X < 0.2<br>Y < 0.4 | 2/4 = 50% | 2/2 = 100% |
| X < 3.75<br>Y < 4.35 | 3/4 = 75% | 3/4 = 75% |
| X < 3.75<br>Y < 4.35<br>Z < 0.6 | 3/4 = 75% | 3/3 = 100% |
| X < 0.5<br>Y < 0.55 | 4/4 = 100% | 4/6 = 67% |

NOTE: Shaded row describes results shown in Figure A.2.

**Figure A.3**
**Classification of Ten Simulations by Vulnerable Conditions**

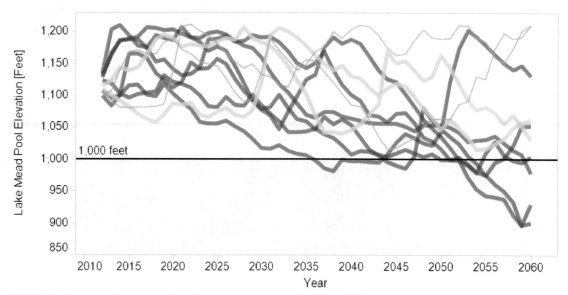

NOTES: Red lines are simulations that do not meet goals and are included in the vulnerable conditions. Yellow lines are those that do meet goals and are included in the vulnerable conditions. Thin gray lines are simulations that meet the Lake Mead goals and are not included in the vulnerable conditions.
RAND *RR242-A.3*

revert to the pre-2007 No Action Alternative, then *Lower Basin Reliability* objectives are not met in most traces. Demand is below this level for three demand scenarios considered in the Basin Study: Rapid Growth C1, Rapid Growth C2, and Enhanced Environment D2 (Figure A.4). Collectively, 85 percent of traces in either of these two sets of vulnerable conditions lead to not meeting objectives, and the conditions capture the large majority (87 percent) of these unacceptable outcomes. However, more than one set of vulnerable conditions is identified in this case, which somewhat reduces interpretability. The vulnerable conditions derived from climate inputs are summarized in Table A.2.

**Figure A.4**
**Second Part of the Low Historical Supply Vulnerable Conditions for Lake Mead Pool Elevation Using Climate Variables**

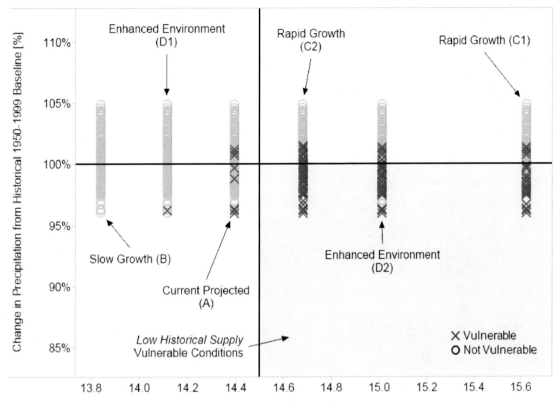

NOTES: Figure only shows results for the Revert to Pre-2007 No Action Alternative scenario and results with temperature trends less than 1 degree Fahrenheit. This scatterplot shows results from thousands of individual traces, many of which overlap.

RAND *RR242-A.4*

**Table A.2**
**Vulnerable Conditions Defined for Lake Mead Pool Elevation with No Continuation of Interim Guidelines**

*Vulnerable Conditions Description: Low Historical Supply, No Continuation of Interim Guidelines*

| Streamflow Conditions | | Climate Conditions | |
|---|---|---|---|
| Vulnerable Traces: 40%<br><br>Vulnerability Statistics:<br>• Coverage: 78%<br>• Density: 68% | Vulnerability Definition:<br>• Revert to No Action Alternative<br>• Annual Mean Natural Flow at Lee Ferry (2012–2060) Less Than 14.6 maf per Year:<br><br>10.0    14.6    18.5<br><br>• Driest Eight-Year Period of Annual Mean Natural Flow at Lee Ferry (2012–2060) Less Than 13.0 maf per Year:<br><br>7.0    13.0    15.5 | Vulnerable Traces: 45%<br><br>Vulnerability Statistics:<br>• Coverage: 93%<br>• Density: 81% | Vulnerability Definition:<br>• Revert to No Action Alternative<br>• Change in Temperature (2012–2060) Greater than 1.7 degrees Fahrenheit<br><br>−0.7  1.7    5.6<br><br>• Average Precipitation (2012–2060) Less Than 100% of Historical Average<br><br>84    100    111<br><br>• 2041–2060 Demand More Than 14.5 maf per year<br><br>13.8    14.5    15.6 |

# Basin Study Options Included in the Portfolios

This appendix provides a brief description for the water-management options evaluated as part of the Basin Study portfolios (Table B.1). Technical Report F (Reclamation, 2012f) provides more detail on all the options evaluated and the qualitative scores developed for each.

**Table B.1**
**Description of Each Type of Option Included in the Portfolio Analysis**

| Option Type | Description | Options Included in Portfolio A (Inclusive) |
|---|---|---|
| Agricultural Conservation | Options that increase water conservation in the agriculture sector and reduce demand for Colorado River water in either the Upper or Lower Basin: Options are disaggregated into 200 taf/year "steps" to represent likely project phasing, and to the Upper and Lower Basins. | • Agricultural Conservation with Transfers (Upper and Lower Basin) |
| Desalination | Options to desalinate (1) ocean water off the California and the Gulf of Mexico coasts, (2) agricultural drainwater, and (3) brackish groundwater: Options are disaggregated into 200 taf/year "steps" to represent likely project phasing. | • Desal–Salton Sea Drainwater<br>• Desal–Pacific Ocean–California<br>• Desal–Gulf |
| Energy Water Use Efficiency | Options to improve the water use efficiency of the energy sector: The conversion of power plants to air cooling calls for the removal of the evaporative cooling systems at the 15 largest power plants in the Basin and installing air-cooling systems. | • Energy Water Use Efficiency–Air Cooling |
| Import | Options to increase the overall water supply of the Basin from other river basins: Imports from the Missouri River and the Mississippi River were considered to augment supply in the Colorado Front Range and reduce the amount of Colorado River exported to these regions. Options for importing to the Green River Headwaters from the Bear, Snake, and Yellowstone Rivers, and to Southern California from Northern sources via water bags and icebergs were also evaluated. Options are disaggregated into 200 taf/year "steps" to represent likely project phasing. | • Import–Front Range–Missouri |
| Local Supply | Local supply options capture local water sources that would otherwise go unused: Rainwater harvesting and Coalbed Methane were the main local options considered. | • Local–Coalbed Methane |
| M&I Conservation | Options that increase water conservation in the municipal and industrial sectors and reduces demand for Colorado River water in either the Upper or Lower Basins: Options are disaggregated into 200 taf/year "steps" to represent likely project phasing. | • M&I Conservation (Upper and Lower Basin) |
| Reuse | Reuse of existing municipal and gray water supplies increase the overall water supply in the Basin: Options are disaggregated into 200 taf/year "steps" to represent likely project phasing. | • Reuse-Municipal<br>• Reuse-Gray Water |

**Table B.1—Continued**

| Option Type | Description | Options Included in Portfolio A (Inclusive) |
|---|---|---|
| Watershed Management | Options that could increase the supply of the Basin by increasing river runoff: Key approaches include Tamarisk control, Forest Management, Brush Control, Dust Control, and Weather Modification. Dust control improves supply by reducing dust accumulation on snow that can lead to earlier snowmelt and more evaporative moisture losses. Weather modification consists of cloud seeding, or adding silver iodide to the atmosphere to serve as condensation nuclei that would increase snowfall over mountain regions. Options are disaggregated into 200 taf/year "steps" to represent likely project phasing. | • Watershed-Weather Modification<br>• Watershed-Dust |
| Groundwater Banking[a] | Option creates an Upper Basin water bank in either Lake Powell or in an offstream groundwater bank to increase protection against curtailment in the Upper Basin: In conjunction with the water bank, various conservation (M&I, agricultural, and energy) efforts across the Upper Basin would be coordinated for the purpose of yielding water to store in the bank. | • Upper Basin Groundwater Bank |

[a] This option is not included in the portfolio option lists. Instead, it is specified to be implemented in all traces for *Portfolio C (Environmental Performance Focus)* and *Portfolio A (Inclusive)*.

# References

Barnett, Tim P., and David W. Pierce, "When Will Lake Mead Go Dry?" *Water Resources Research*, Vol. 44, No. 3, 2008, p. W03201.

———, "Sustainable Water Deliveries from the Colorado River in a Changing Climate," *Proceedings of the National Academy of Sciences*, Vol. 106, No. 18, 2009, pp. 7334–7338.

Barnett, Tim P., David W. Pierce, Hugo G. Hidalgo, Celine Bonfils, Benjamin D. Santer, Tapash Day, Govindasamy Bala, Andrew W. Wood, Toru Nozawa, Arthur A. Mirin, Daniel R. Cayan, and Michael D. Dettinger, "Human-Induced Changes in the Hydrology of the Western United States," *Science*, Vol. 319, No. 5866, 2008, pp. 1080–1083. As of February 1, 2010:
http://search.ebscohost.com/login.aspx?direct=true&db=a9h&AN=31205697&site=ehost-live

Bryant, Benjamin P., and Robert J. Lempert, "Thinking Inside the Box: A Participatory, Computer-Assisted Approach to Scenario Discovery," *Technological Forecasting and Social Change*, Vol. 77, No. 1, 2010, pp. 34–49.

Carpe Diem West Academy, *Roadmap: Decision Making Framework* web page, 2013. As of March 15, 2013:
http://carpediemwestacademy.org/roadmap

Christensen, Niklas S., and Dennis P. Lettenmaier, "A Multimodel Ensemble Approach to Assessment of Climate Change Impacts on the Hydrology and Water Resources of the Colorado River Basin," *Hydrology and Earth System Sciences*, Vol. 11, No. 4, 2007, pp. 1417–1434.

Christensen, Niklas S., Andrew W. Wood, Nathalie Voisin, Dennis P. Lettenmaier, and Richard N. Palmer, "The Effects of Climate Change on the Hydrology and Water Resources of the Colorado River Basin," *Climatic Change*, Vol. 62, No. 1–3, 2004, pp. 337–363.

Dixon, Lloyd, Robert J. Lempert, Tom LaTourette, and Robert T. Reville, *The Federal Role in Terrorism Insurance*, Santa Monica, Calif.: RAND Corporation, MG-679-CTRMP, 2007. As of August 5, 2013:
http://www.rand.org/pubs/monographs/MG679.html

Fischbach, Jordan R., David R. Johnson, David S. Ortiz, Benjamin P. Bryant, Matthew Hoover, and Jordan Ostwald, *Coastal Louisiana Risk Assessment Model*, Santa Monica, Calif.: RAND Corporation, TR-1259-CPRA, 2012. As of August 4, 2013:
http://www.rand.org/pubs/technical_reports/TR1259.html

Friedman, Jerome H., and Nicholas I. Fisher, "Bump Hunting in High-Dimensional Data," *Statistics and Computing*, Vol. 9, No. 2, 1999, pp. 123–143.

Glennon, Robert, and Michael J. Pearce, "Transferring Mainstream Colorado River Water Rights: The Arizona Experience," *Arizona Law Review*, Vol. 49, 2007, pp. 235–256.

Groves, David G., and Robert J. Lempert, "A New Analytic Method for Finding Policy-Relevant Scenarios," *Global Environmental Change Part A: Human and Policy Dimensions*, Vol. 17, No. 1, 2007, pp. 73–85.

Hoerling, M., and J. Eischeid, "Past Peak Water in the Southwest," *Southwest Hydrology*, Vol. 6, No. 1, 2006.

Lempert, Robert J., and Myles T. Collins, "Managing the Risk of Uncertain Threshold Responses: Comparison of Robust, Optimum, and Precautionary Approaches," *Risk Analysis: An International Journal*, Vol. 27, No. 4, 2007, pp. 1009–1026.

Lempert, Robert J., and David G. Groves, "Identifying and Evaluating Robust Adaptive Policy Responses to Climate Change for Water Management Agencies in the American West," *Technological Forecasting and Social Change,* Vol. 77, 2010, pp. 960–974.

Lempert, Robert, Nidhi Kalra, Suzanne Peyraud, Zhimin Mao, Sinh Bach Tan, Dean Cira, and Alexander Lotsch, *Ensuring Robust Flood Risk Management in Ho Chi Minh City,* Policy Research Working Paper, No. WPS 6465, Washington, D.C.: The World Bank, May 2013. As of August 5, 2013: http://documents.worldbank.org/curated/en/2013/05/17784406/ ensuring-robust-flood-risk-management-ho-chi-minh-city

Lempert, Robert J., and Steven W. Popper, "High-Performance Government in an Uncertain World," in Robert Klitgaard and Paul C. Light, eds., *High Performance Government: Structure, Leadership, Incentives,* Santa Monica, Calif.: RAND Corporation, MG-256-PRGS, 2005. As of May 7, 2012: http://www.rand.org/pubs/monographs/MG256.html

Lempert, Robert J., Steven W. Popper, and Steven C. Bankes, *Shaping the Next One Hundred Years: New Methods for Quantitative, Long-Term Policy Analysis,* Santa Monica, Calif.: RAND Corporation, MR-1626-RPC, 2003. As of May 4, 2012: http://www.rand.org/pubs/monograph_reports/MR1626

MacDonnell, Lawrence J., David H. Getches, and William C. Hugenberg, Jr., "The Law of the Colorado River: Coping with Severe Sustained Drought," *Journal of the American Water Resources Association,* Vol. 31, 1995, pp. 825–836.

Means, Edward, Maryline Laugier, Jennifer Daw, Laurna Kaatz, and Marc Waage, *Decision Support Planning Methods: Incorporating Climate Change Uncertainties into Water Planning,* San Francisco, Calif.: Water Utility Climate Alliance, 2010. As of May 7, 2012: http://www.wucaonline.org/assets/pdf/actions_whitepaper_012110.pdf

Meko, David M., Connie A. Woodhouse, Christopher A. Baisan, Troy Knight, Jeffrey J. Lukas, Malcolm K. Hughes, and Matthew W. Salzer, "Medieval Drought in the Upper Colorado River Basin," *Geophysical Research Letters,* Vol. 34, No. 10, 2007, p. L10705.

Milly, P.C.D., Julio Betancourt, Malin Falkenmark, Robert M. Hirsch, Zbigniew W. Kundzewicz, Dennis P. Lettenmaier, and Ronald J. Stouffer, "Stationarity Is Dead: Whither Water Management?" *Science,* Vol. 319, No. 5863, 2008, pp. 573–574.

Nash, Linda L., and Peter H. Gleick, *The Colorado River Basin and Climate Change: The Sensitivity of Streamflow and Water Supply to Variations in Temperature and Precipitation,* Washington, D.C.: U.S. Environmental Protection Agency, Vol. 121, 1993.

National Academies of Science: Committee on America's Climate Choices, *America's Climate Choices,* Washington, D.C.: The National Academies Press, 2011. As of March 15, 2013: http://www.nap.edu/openbook.php?record_id=12781

National Research Council, *Colorado River Basin Water Management: Evaluating and Adjusting to Hydroclimatic Variability,* National Academies Press, 2007.

———, *Informing Decisions in a Changing Climate,* Washington, D.C.: National Academies Press, 2009. As of April 30, 2012: http://www.nap.edu/openbook.php?record_id=12626

Popper, Steven W., Claude Berrebi, James Griffin, Thomas Light, Endy M. Daehner, and Keith Crane, *Natural Gas and Israel's Energy Future: Near-Term Decisions from a Strategic Perspective,* Santa Monica, Calif.: RAND Corporation, MG-927-YSNFF, 2009. As of August 5, 2013: http://www.rand.org/pubs/monographs/MG927.html

Public Law 111-11, Title IX, Bureau of Reclamation Authorizations, Subtitle F, Secure Water, March 30, 2009.

Rajagopalan, Balaji, Kenneth Nowak, James Prairie, Martin Hoerling, Joseph Barsugli, Andrea Ray, and Bradley Udall, "Water Supply Risk on the Colorado River: Can Management Mitigate?" *Water Resources Research,* Vol. 45, 2009.

RAND, RDMlab website, undated. As of October 2013:
http://www.rand.org/rdmlab

Reclamation—*See* U.S. Bureau of Reclamation.

Seager, Richard, Ting Mingfang, Isaac Held, Vochanan Kushnir, Lu Jian, Gabriel Vecchi, Huang Huei-Ping, Nih Harnik, Ants Leetmaa, Lau Ngar-Cheung, Li Cuihua, Jennifer Velez, and Naomi Naik, "Model Projections of an Imminent Transition to a More Arid Climate in Southwestern North America," *Science*, Vol. 316, No. 5828, 2007, pp. 1181–1184.

Seager, Richard, Mingfang Ting, Cuihua Li, Naomi Naik, Ben Cook, Jennifer Nakamura, and Haibo Liu, "Projections of Declining Surface-Water Availability for the Southwestern United States," *Nature Climate Change*, December 23, 2012. As of August 5, 2013:
http://www.nature.com/nclimate/journal/v3/n5/full/nclimate1787.html

Simon, Herbert A., "Rational Choice and the Structure of the Environment," *Psychological Review*, Vol. 63, No. 2, 1956, pp. 129–138.

Stockton, C. W., and G. C. Jacoby, "Long-Term Surface-Water Supply and Streamflow Trends in the Upper Colorado River Basin," *Lake Powell Research Project Bulletin*, Vol. 18, No. 18, 1976, p. 70.

United States, Mexico, Treaty on Utilization of Waters of the Colorado and Tijuana Rivers and of the Rio Grande, 1944.

U.S. Bureau of Reclamation, *DRAFT Annual Operating Plan for Colorado River Reservoirs, 2014*, U.S. Department of Interior, 2013. As of September 16, 2013:
http://www.usbr.gov/lc/region/g4000/AOP2014/AOP14_draft.pdf

———, *Colorado River Compact*, 1922. Accessed on Reclamation website, March 21, 2013:
http://www.usbr.gov/lc/region/g1000/pdfiles/crcompct.pdf

———, *Colorado River Interim Guidelines for Lower Basin Shortages and Coordinated Operations for Lakes Powell and Mead—Final Environmental Impact Statement*, U.S. Department of the Interior, November, 2007.

———, *Colorado River Basin Water Supply and Demand Study Technical Report A: Scenario Development*, U.S. Department of the Interior, 2012a. As of December 31, 2012:
http://www.usbr.gov/lc/region/programs/crbstudy/finalreport/Technical%20Report%20A%20-%20
Scenario%20Development/TR-A_Scenario_Development_FINAL_Dec2012.pdf

———, *Colorado River Basin Water Supply and Demand Study Technical Report B: Water Supply Assessment*, U.S. Department of the Interior, 2012b. As of December 31, 2012:
http://www.usbr.gov/lc/region/programs/crbstudy/finalreport/Technical%20Report%20B%20-%20Water%20
Supply%20Assessment/TR-B_Water_Supply_Assessment_FINAL_Dec2012.pdf

———, *Colorado River Basin Water Supply and Demand Study Technical Report C: Water Demand Assessment*, U.S. Department of the Interior, 2012c. As of December 31, 2012:
http://www.usbr.gov/lc/region/programs/crbstudy/finalreport/Technical%20Report%20C%20-%20
Water%20Demand%20Assessment/TR-C_Water_Demand_Assesmemt_FINAL_Dec2012.pdf

———, *Colorado River Basin Water Supply and Demand Study Technical Report D: System Reliability Metrics*, U.S. Department of the Interior, 2012d. As of December 31 2012:
http://www.usbr.gov/lc/region/programs/crbstudy/finalreport/Technical%20Report%20D%20-%20
System%20Reliability%20Metrics/TR-D_Syst_Reliab_Metrics_FINAL_Dec2012.pdf

———, *Colorado River Basin Water Supply and Demand Study Technical Report E: Approach to Develop and Evaluate Options and Strategies*, U.S. Department of the Interior, 2012e. As of December 31, 2012:
http://www.usbr.gov/lc/region/programs/crbstudy/finalreport/Technical%20Report%20E%20-%20
Approach%20to%20Develop%20and%20Evaluate%20Options%20and%20Strategies/TR-E_Approach_to_
Eval_Ops_and_Strats_FINAL_Dec2012.pdf

———, *Colorado River Basin Water Supply and Demand Study Technical Report F: Development of Options and Strategies*, U.S. Department of the Interior, 2012f. As of December 31, 2012:
http://www.usbr.gov/lc/region/programs/crbstudy/finalreport/Technical%20Report%20F%20-%20
Development%20of%20Options%20and%20Stategies/TR-F_Devlpmnt_of_Ops_and_Strats_FINAL_
Dec2012.pdf

———, *Colorado River Basin Water Supply and Demand Study Technical Report G: System Reliability Analysis and Evaluation of Options and Strategies*, U.S. Department of the Interior, 2012g. As of December 31, 2012: http://www.usbr.gov/lc/region/programs/crbstudy/finalreport/Technical%20Report%20G%20-%20 System%20Reliability%20Analysis%20and%20Evaluation%20of%20Options%20and%20Stategies/TR-G_ System_Reliability_Analysis_FINAL_Dec2012.pdf

———, *Colorado River Basin Water Supply and Demand Study: Study Report*, U.S. Department of the Interior, December, 2012h. As of February 15, 2013: http://www.usbr.gov/lc/region/programs/crbstudy/finalreport/Study%20Report/StudyReport_FINAL_ Dec2012.pdf

Wildman, Richard A., and Noelani A. Forde, "Management of Water Shortage in the Colorado River Basin: Evaluating Current Policy and the Viability of Interstate Water Trading 1," *Journal of the American Water Resources Association*, 2012.

Woodhouse, Connie A., "A 431-Year Reconstruction of Western Colorado Snowpack from Tree Rings," *Journal of Climate*, Vol. 16, 2003, pp. 1551–1561.

Woodhouse, C. A., S. T. Gray, and D. M. Meko, "Updated Streamflow Reconstructions for the Upper Colorado River Basin," *Water Resources Research*, Vol. 42, 2006, p. W05415.

Zagona, Edith A., Terrance J. Fulp, Richard Shane, Timothy Magee, and H. Morgan Goranflo, "RiverWare: A Generalized Tool for Complex Reservoir System Modeling," *Journal of the American Water Resources Association*, Vol. 37, No. 4, 2001, pp. 913–929.